[A Naturalist's Gui]de to the [BUTTERF]LIES OF INDIA

Pakistan, Nepal, Bhutan, Bangladesh and Sri Lanka

Peter Smetacek

JOHN BEAUFOY PUBLISHING

First published in the United Kingdom in 2017 by John Beaufoy Publishing Ltd
11 Blenheim Court, 316 Woodstock Road, Oxford OX2 7NS, England
www.johnbeaufoy.com

10 9 8 7 6 5 4 3 2 1

Photo Credits
Front Cover: *clockwise from top right* Yellow Gorgon (Ngangom Aomoa); Common Blue Apollo (Peter Smetacek); Popinjay (Ngangom Aomoa); Southern Blue Oakleaf (Kishen Das); Pale Himalayan Oakblue (Peter Smetacek);
Back Cover: Malabar Banded Peacock (C. Susanth). **Title Page:** Orange Oakleaf (Ngangom Aomoa).
Contents Page: China Nawab (Alka Vaidya)

Photographs are denoted by a page number, a row number from the top in brackets and L = left, C= centre, ML = Middle Left, MR = Middle Right and R = right.)
Introduction: 5 [1L] Tawny Coster; [1R] Paris Peacock (Binita Goswami); [2] Common Tiger
14 [1L] Mottled Emigrant (1R) Himalayan Yellow Banded Flat; [2L] Silver Hairstreak (2R) Yellow Pansy; [3L] Mixed Punch female; (3R) Lime (Milind Bhakare);

Main descriptions: see page 156 and 157

The author would like thank Rajashree Bhuyan for editorial help.

ISBN 978-1-909612-79-2

Edited by Krystyna Mayer
Designed by Gulmohur India

Printed and bound in Malaysia by Times Offset (M) Sdn. Bhd.

Contents

INTRODUCTION

The Indian subcontinent contains representations of every major climate and flora found on the globe, from tropical evergreen forest, mangroves, tropical deciduous forest, grassland, scrubland, hot desert in the Thar and cold desert in Ladakh, to temperate deciduous forest and polar vegetation in the Himalaya. This favourable situation supports a vast variety of fauna; indeed, India is home to the greatest variety of large cats, bears, deer and monkeys – and 1,318 butterfly species. Pakistan is home to some 450 butterfly species, Sri Lanka to 245 and Nepal to 640, and work is still underway to determine the fauna of Bhutan and Bangladesh.

GEOGRAPHY AND CLIMATE

The Indian subcontinent is a unique geological formation, protected from the north by the Himalaya, open in the south to the monsoon winds, and straddling an equable zone between the western Palaearctic and the eastern tropics. Minor geological formations on the subcontinent have influenced local climate, so that the coasts are flanked by the Eastern Ghats and the Western Ghats. The seaward face of these hills receives heavy rainfall and supports tropical evergreen forest, while the leeward side supports tropical deciduous vegetation. Further north, the Gangetic plain drains the precipitation falling on the Himalaya and the northern parts of the Deccan Plateau. The western borders of the subcontinent are covered by desert. However, this desert appears to be of recent origin, for accounts dating to the invasion of India by Alexander the Great 2,300 years ago describe dense forest covering the landscape in that area.

INSECTS

Insects comprise the largest group of life forms on land, both in terms of number of individuals and in number of different species. Their diminutive size renders most of them inconspicuous to the casual eye and certainly, most of them go to great lengths to remain, as far as possible, unnoticed as a means of survival. Insects comprise the staple diet of many vertebrates – in fact, a large group of vertebrates is called insectivores. There are also many insects that prey on other insects, besides fungus, bacteria and even some plants.

Historically, little attention was paid to these pervasive creatures. The task of studying and understanding them was taken up seriously only after the eighteenth century, when the flourishing sea trade sparked European interest in creatures from far-flung corners of the globe, and when means of classifying them and storing them properly were devised.

The classification of animals (including insects) is a complex issue, with different authorities often having different views on the matter. However, it basically involves grouping animals into hierarchies, from the most complex to the simplest. The development of a hierarchy was devised in the eighteenth century by the Swedish naturalist Carl Linnaeus, who proceeded to name many insects, mammals, birds and plants. Insects are grouped in the class Insecta, while the neighbouring class, Arthropoda, contains eight-legged creatures such as scorpions and spiders.

Butterfly and moth classification

Butterflies and moths are insects that belong to the order Lepidoptera, from the Greek, 'scale' + 'wing', referring to the tiny, coloured, tile-like 'scales' covering most of their wing area. These scales, which come away like dust on the hand, are arranged neatly over the transparent wing membranes, and are responsible for the colours and patterns on the wings. There are several scale types, some in the form of hairs, some like the scales on a

PARTS OF A BUTTERFLY

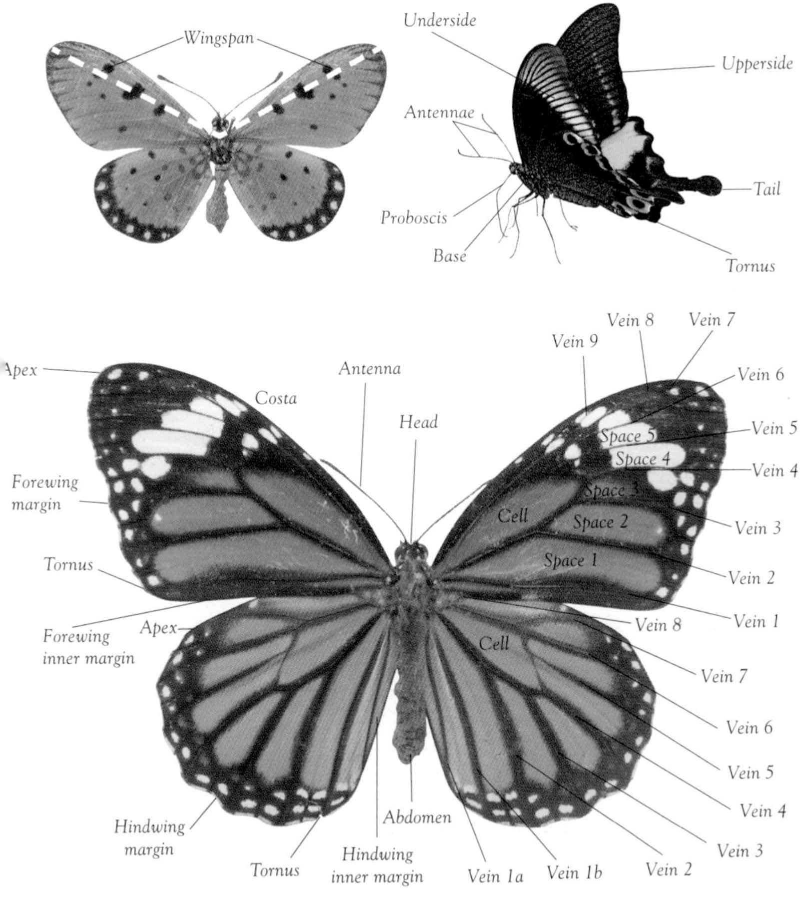

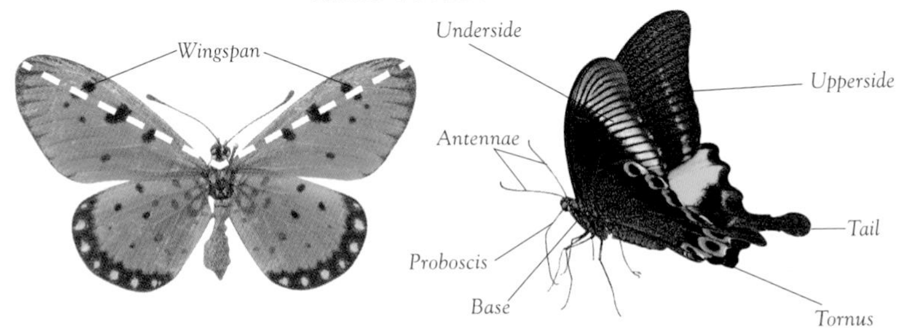

fish and others like roof tiles. Then there are pigment scales and iridescent scales. Pigment scales, as the name implies, are coloured scales that make up the pattern. Iridescent scales are the scales responsible for the fascinating flashes of colour, usually blues and greens, on the wings of some tropical butterflies, such as the blue crows. These scales are rather remarkable in that they are, on their own, colourless, but structurally break down incoming white sunlight so as to reflect those brilliant flashes of light. Viewed from other angles their wings appear dull, usually brown.

Although the distinction of this order of creatures is implicit in the name, butterflies and moths possess other distinctive characteristics such as the life cycle, which proceeds in four clearly defined stages, and the inability of all members (except one primitive family) to eat solid food during the adult stage, that is, in the stage when the creature is a butterfly or moth. Within these limits, there is every logically conceivable variation, from certain wingless moths to butterflies and moths that, in the caterpillar stage, feed on aphids, ripening fruits and rotting wood, to a species of moth that does not mind an occasional sip of mammalian blood to wash down its usual diet of fruit juice.

The distinction between butterflies and moths is not at all clear and is more of an artificial one than a natural one. Currently, butterflies are included in one of the superfamilies of the Lepidoptera, the Papilionoidea. Butterflies, as a group, possess certain distinctive physiological similarities, such as a bulge, or 'club', at the end of each antenna, a tube-like mouth or probosis, a different structure of veins or struts on the forewings and hindwings from that of moths, and the fact that both sets of wings are coupled by the overlapping of a small, stiffened area on both wings, which is specially strengthened in the area of overlap. This last distinction was included to separate what we call butterflies from the day-flying moth family Castniidae, members of which also have clubbed antennae, but have the forewing and hindwing coupled with a stiff bristle (frenulum) arising from near the base of the hindwing, which fits into a flap on the forewing (retinaculum). The exception to the generalities about butterflies is a small group of 'primitive' Australian Hesperiidae (skippers), the wings of which are coupled by a rudimentary frenulum, and the African Pierid *Pseudopontia paradoxa*, in which the antennae do not end in a swelling.

An important distinction arises out of the fact that most butterflies depend on the sun's energy to warm them enough to fly, while most moths have a system of warming themselves by vibrating their wings rapidly.

'Moths' is a very loose term, which includes everything that is a Lepidoptera but not a butterfly. To give a rough idea of proportion, there are at least ten times more moths than all butterflies, and this difference will probably grow as moths are better investigated.

The butterfly superfamily is divided into families, on the basis of certain physiological characteristics. Family is a rather broad level of classification, and it is narrowed down into subfamilies, grouped on the basis of arbitrarily assigned similarities. These are divided into genera (singular genus), or groups distinguished on the basis of similar physical structures or some such criteria. A genus may comprise one or several butterfly species. Butterfly species are recognized on the basis of reproductive isolation, that is, the inability of opposite sexes of different species to beget fertile offspring.

All of what are today known as 'species' developed, not necessarily from a common 'genus' or 'family' level ancestor, but by responding differently to environmental factors so that, in time, different populations exploited and came to depend upon different environmental factors, and adjusted themselves accordingly. During the course of this process, genetic compatibility and other factors governing reproduction became different enough to disallow cross-fertilization between populations.

It should be stressed that the classification of everything above the species level is merely our 'map' for studying the 'territory' of butterflies. There are no doubt several members of genera and even subfamilies that have most probably evolved from an ancestral species but, lacking concrete data and evidence of environmental factors that caused them, such relationships are at present no more than speculation.

Having arrived at the understanding that butterfly, moth, order, family and genus are merely words and groups established to help us communicate among ourselves about these creatures, the species demand attention. To use an analogy, dogs are creatures most of us are familiar with. Although there are striking differences between different breeds, as between terriers and mastiffs, all domestic dogs are grouped together under one species, *Canis lupus*, on the basis of their ability to reproduce and produce potentially fertile mongrels. Similarly, a butterfly species is capable of great variation in size, colour and pattern in response to seasons or other environmental factors. Seasonal forms are generally referred to as wet-season form (WSF) and dry-season form (DSF), though the terms 'dry' and 'wet' have been chosen on very general considerations. If a population of a species differs somewhat from area to area or due to other factors, it is referred to as a subspecies, or geographical variation. To continue with the analogy referred to above, relating to dogs, the domestic dog is a divergent subspecies, *Canis lupus familiaris*, of the wolf, produced in this case by selective breeding by humans; the name '*familiaris*' indicates the name of the subspecies.

Another distinction arises when some of the adult butterflies are markedly different from others, though they were born of the same parents. When such a phenomenon occurs regularly, these aberrants are called forms. Some Indian butterflies, such as the Common Mormon *Papilio polytes*, are capable of producing three markedly different female butterfly types, but only one type of male from the same batch of eggs.

Summing up the above, each butterfly species is usually referred to by two names, the first indicating the genus to which it belongs and the second the species. Therefore, *Papilio polytes* indicates that the species *polytes* as distinguished from all other butterflies belongs to the genus *Papilio*. *Papilio polytes polytes* refers to the subspecies *polytes* of the species *Papilio polytes*. *Papilio polytes* f. (form) *theseus* refers to the form *theseus* of *polytes*. Since this variation is restricted to the females of the species, it generally goes without saying that the butterfly under consideration is a female.

In this system, names of butterflies are usually derived from Latin or Greek. The pronunciation of some names might look challenging to the beginner, but an understanding of Latin pronunciation, which is more similar to Hindi and German than to English, makes it easier to remember 'complicated' names. The letter 'a' is usually pronunced as in 'star'. The letter 'y' derives from the Greek ipsilon, and is pronounced rather like the 'i' in 'inch'. In older literature, the letter 'j' is pronounced as the English 'y' (as in *lajus* and *jarbus*).

Although there are many species in which there is a considerable superficial difference between the sexes, males and females look alike in perhaps an equal number of cases. Lepidopterists tell them apart by examining the genitalia close up. Females are generally larger, fly slower and have more rounded wings than males.

Metamorphosis in butterflies

EGG

All butterflies begin life as an egg. This varies in shape, colour and size according to the species, and it is usually possible to identify the species or at least the group on the basis of the egg.

Once the little caterpillar (or larva) has been fashioned within the egg, a process that might take one to several weeks, the eggshell cracks at a certain point, which is usually structurally weaker than the other parts of the shell.

CATERPILLAR

The first task of the caterpillar is to hungrily devour the egg-case, after which it proceeds to search for its food. This can be anything from the leaves of a particular type of plant, to certain flower buds, fruits, aphids or, in the case of some moths, dry wood. The female butterfly usually lays its eggs on or very near the larval food plant, so the caterpillar does not need to search far for food. Its main aim is to gorge itself until it is a considerable number of times larger than when it emerged from the egg.

As the body of the caterpillar enlarges, it sheds its outer skin when it is incapable of expanding any more. This happens 3–5 times during the caterpillar stage, and the process is called moulting. The interesting feature of moulting is that after a moult, the new skin (and consequently the caterpillar) may, in some species, assume quite a different colour, pattern and, in some cases, shape. It may be difficult to recognize the yellow-green caterpillar that you saw yesterday as the same one that has became a sedate brown one, until you find a crumpled little green skin in the vicinity. Most of the nutrients are drawn back into the body before the old skin is discarded, and in some cases the caterpillar devours the cast-off skin.

The majority of caterpillars feed on the leaves of plants. The freshly emerged caterpillars first eat the young reddish-green leaves of their food plant, mainly because these leaves are easier to chew, but also because they have a high nutrient content. Once its jaws have developed to the extent that it can chew 'harder' stuff, a caterpillar moves on to the mature leaves. Caterpillars that eat dry wood and other forest debris take rather longer than leaf-eating caterpillars to develop. This is because the green leaves of a plant usually contain the highest amount of nutrients, followed by twigs, roots and, finally, woody parts.

The caterpillar, chewing away at the leaves, grows at a remarkable rate, converting a large part of the food it extracts from leaves into fatty tissue. Once it has reached a level of satiation, it grows sluggish and crawls off to find a relatively safe place to pupate. The larval stage lasts for 2–8 weeks, depending upon the availability of food.

Certain blues (Lycaenidae) such as *Spindasis* (silverlines), *Surendra* (acacia blues) and *Arhopala* (oakblues), are attended by ants during the caterpillar stage. The female butterfly lays its eggs near foraging ants of certain 'friendly' species. The ants take care of the egg and the caterpillar that emerges, guarding, cleaning and in some cases bringing food for the honoured guest. In return, the well-protected, sluggish guest secretes a sweet liquid from special glands on the rear half of its back. The ants devour this secretion and ask for nothing else in return. In some cases, the caterpillar lives and pupates within the ants' nest.

PUPA

A few butterflies build a rough coocon, that is, an outer shell of silken thread or other material. Generally, they simply attach themselves to a twig or other firm support, sticking themselves onto it at their rear end, and wrapping a neat silken girdle around their support and themselves. Once attached, they shed their final larval skin. A completely different form reveals itself when the larval skin slips away. The pupation stage lasts for between two weeks to several months (in case the species has to pass the winter in this stage). In some moths, it may lasts for 2–3 years as the moth inside the pupa waits for what it feels is the right time to emerge.

Although this looks as dormant a stage as the egg, the outward lack of activity is misleading. Within the walls of the pupa, the caterpillar has, except for a few vital organs, dissolved into a liquid and its cells rearrange themselves into an insect that can fly.

BUTTERFLY

Once the butterfly is ready, it waits for the right moment to crack the stiff pupal case and emerge from a sheath for the last time in its life. When it judges conditions to be favourable for its emergence, plasma 'blood' rushes to the brain and tissue of the insect, galvanizing it into action. The pupal case is split and the butterfly crawls out and proceeds to expand its wings, which consist of four stubs on its back. The stubs unfold to reveal large wings comprising a system of hollow veins covered with membranes on both surfaces. The transparent membranes are covered with colourful scales, which give the Lepidoptera their name. The wings expand as fluid is pumped into them. The pressure of the fluid causes the wings to straighten, much in the manner inflatable toys assume their shape when pumped up. Once the wings have expanded to the fullest extent, the butterfly proceeds to dry them in the sun.

Butterflies generally emerge from the pupal case during pre-dawn darkness or at dawn, using the early-morning sunlight to dry their wings. They find a place which they can hang from, except some species of the Hesperiiidae, which expand their wings and dry them held upright above their bodies. The wet wings are initially held horizontally open, perpendicular to the sun's rays. When they are dry and rigid, the butterfly flexes the wing muscle several times, opening and shutting the wings. Then, with complete confidence, it launches itself into the air and exhibits, from the very beginning, all the flying skills of its predecessors.

In this stage, as far as is understood, a butterfly's main task is reproduction, and for the females, dispersion of the species. However, flying is an energy-expensive mode of travel. Most butterflies supplement their body-fat fuel reservoirs held over from the larval stage with energy-rich nectar from flowers and sap from fruits or trees. On hot days in the tropics, some butterfly species congregates at patches of wet mud and have been observed to 'flush' their system, greedily sucking in moisture and ejecting quantities of liquid from their anal opening. This and the predilection of certain species for bird or mammal droppings is believed to be a method of absorbing mineral salts. However, most butterfly species, on emergence from the pupal case, are capable of reproducing and laying eggs without the intake of any food.

THE SENSES

Butterflies are known to have faculties that enable them to taste, touch, see, smell and respond to sound. The palpi, a pair of appendages on either side of the proboscis, or mouth tube, as well as the forelegs, are thought to contain organs of taste. The large compound eyes are responsible for sight, while the antennae (feelers) are thought to be responsible for the sense of smell as well as balance, since a butterfly deprived of these makes very irregular progress, bumping into trees and other obstacles. Organs for the reception of sound waves are situated along the body. It should not be assumed that these organs interpret sound waves in exactly the same manner as we do.

Butterflies have an acutely developed sense of chemical detection. A sexually mature adult, male or female, signals this fact to the opposite sex by dissipating a drop or two of a chemical compound, called a pheromone, which enables the opposite sex to home in on the individual that released the pheromone, in some cases perhaps over a distance of a kilometre.

It is uncertain whether butterflies posses other senses in addition to the above.

ADAPTATION IN BUTTERFLIES

Adaptation is associated with the capability of a population or species to modify itself according to prevailing circumstances. A healthy population of a species may be said to represent the validity of the information package represented by the species in relation to its chosen environment. Conversely, a rapidly dwindling population suggests that the information represented by the species is currently invalid in relation to its environment. At the root of the proliferation or extinction of a species lies its ability to change its habits, dependence patterns, physical structure or requirements.

This ability is manifested in ingenious ways. In butterflies, the most intriguing is the phenomenon known as mimicry, or the endeavour to look like something else. The most famous examples probably occur in oakleaf butterflies, which imitate, almost to perfection, dry leaves of trees found in their chosen habitat. In other species, the undersides of the wings are patterned to blend against the bark of a tree or forest debris.

Yet other butterflies, such as satyrines, have developed fierce-looking eyespots on their wings. It is believed that in some cases, these are meant to resemble the eyes of something much larger and thus startle a predator stalking or attacking the butterfly. In other cases, the eyespots serve to divert attack from a sensitive part of the butterfly, such as the body, to a disposable part of the wings, since predators usually attack eyespots in the belief that they are attacking the head of their prey.

There are several types of inter-butterfly mimicry, which are referred to by the names of the naturalists who studied and described them. An example is where butterflies that contain poisonous substances in the body and are thereby 'protected' from most predators, develop similar wing patterns and habits, in order to reduce the cost of advertising their unpalatablilty to naive new generations of predators. This trait was observed and recorded by Fritz Müller and is known as Müllerian mimicry. It is apparent in groups such as the windmills (*Byasa* group), batwings (*Atrophaneura* group), and several danaines, such as crows (*Euploea core* group), and glassy tigers (*Parantica* group).

Another type of mimicry, which was studied by William Bates, involves a level of deception, where palatable butterflies mimic the wing patterns and habits of poisonous ones and thereby survive. This is known as Batesian mimicry.

The unit of survival in nature is a population along with its chosen environment, that is, those surroundings where the population is capable of adapting to extremes, whether predator pressure, climatic fluctuations, outbreaks of disease or competition from other species dependent on the same food source.

The phenomenon of melanism, that is, the blackening of the overall shade on the wings of butterflies and moths, is genetically controlled. In butterflies, melanism is a response to climate. It is generally true that butterflies of the dry western Himalaya are paler than their eastern Himalaya counterparts, while butterflies of the dry western plains are paler than populations from the southern Indian evergreen forests.

Butterfly races are also genetically controlled. In the case of *Papilio polytes*, where three different types of female are capable of emerging from the same batch of eggs, there is a predominance of those that resembles local models in the area. The female form *romulus* of this butterfly looks very much like *Pachliopta hector* and only occurs in the same localities as the latter, while the form *stichius* looks like and mimics *Pachliopta aristolochiae*, *P. polyphontes* and *P. polydorus* in different parts of its range. *P. polydorus*, which occurs east of Sulawesi and south of the Philippines, is tailless, and the mimicking form *polytes* of *Papilio polytes* in those areas has very short tails on the hindwings.

SIGNIFICANCE OF ADAPTATION

The variety of sizes, shapes, colours and patterns of butterflies has developed so as to enable them to survive in their natural habitats. In the tropics, a single locality can contain several hundred species. Diversity is caused by differences in genetic architecture, wherein is encoded the information relevant to the species. Thus, each species in nature is, to a large extent, a reflection of a causative factor in the environment, the response to which by an organism can, in some cases, give valuable insight into the requirements of the species.

At different seasons, different species of butterfly tend to dominate the community – in other words, certain species prefer to emerge from their pupae during certain seasons, when conditions are most favourable to them. This dominance of the community usually lasts for a few weeks, rarely for several months.

The rate at which insects can reproduce is proverbial, but the point to be derived from this is that each generation of insects is theoretically capable of mutating to the limit of its genetic capability and, once a certain change has been effected and incorporated by succeeding generations, they are yet again in position to take an evolutionary leap through genetic changes. Thus, the stability-instability-adjustment-new level of stability cycle can take place within the space of a few generations. The African butterfly *Bicyclus anynana* was selectively bred under laboratory conditions to change from brown to purple, a change that was managed in six generations in the course of a year. Given that most insect species raise at least two generations annually and that some, such as Houseflies, are capable of reproducing continuously under favourable conditions, it is hardly surprising to see why crop-pest control technology turned into an arms race, with approximately 30 per cent of the global harvest still being lost to species considered to be pests despite the vast amount of pesticides produced and used to try and control them.

Each new generation is potentially capable of altering its physiological characteristics to deal with prevailing environmental changes or constraints, as is exemplified in the striking difference between wet- and dry-season forms in Indian and other tropical butterflies.

The ability to rapidly adjust to environmental changes has led to insect colonization of almost the entire land area, except perpetually icebound regions. This constraint is not due to an inability of insects to adjust to such conditions, but rather to the inability of land plants, on which the insects depend, to adjust to these conditions.

Some insects have used this ability to mutate in order to exploit specific niches in their chosen habitat, such as the ability to dodge at rather high speed through dense bamboo undergrowth, developed by the genera *Lethe* and *Mycalesis*. All species that depend on 'cryptic colouration', for example, are bound to environment areas where their colouration is relevant. Outside such areas, their unusual colouration might even draw attention to them. Hence, highly specialized species are not usually found outside localities preferred by them. Such specialization is found in the leaf butterflies (*Kallima*, *Doleschallia* and *Amblypodia*).

The opposite of this logic of exploiting details of an environment for survival of the species is practised by some other groups that have adjusted to different factors, so that some species are able to survive in dissimilar environments. This strain is exemplified by migratory butterflies such as the Painted Lady *Vanessa cardui* and Peablue *Lampides boeticus*. The habitat of such species may be described as 'cosmopolitan'.

These two broadly opposite lines of evolution are in no way clearly demarcated – in fact, they tend to overlap in the case of certain factors and are intended only as a provisional 'map' for examining the significance of the habits and colouration of butterflies.

The factors that a butterfly genetically adjusts to include the most diverse details and generalities of its preferred environment. The 'specialized' stream is bound by factors such as the quantity and quality of shade, dependence on specific partners such as certain ants or plants, quantity and type of forest litter, or density of secondary growth in a forest.

Members of the 'generalized' stream have adjusted by:
- Developing the ability to feed on a variety of plants in the larval stage.
- Developing wing and body aerodynamics suitable for long flights so as to be able to exploit the potential of several different localities.
- Developing protective devices such as unpalatability to a wide range of insectivorous animals.
- Affecting fast, erratic flight in the adult stage to avoid predators.

The process involved in a butterfly species' choice of larval food plant – though not fully understood – deserves examination.

At two stages of its life, that is the egg and pupal stages, the insect is reduced to a liquid form, while structural changes are wrought by a chemical programme. At both these stages, there is no waste product, except in a gaseous form and in some cases a drop of liquid left in the pupal case after the butterfly has emerged. Nor is there any intake of nutrients. Therefore, to a large extent, these two stages characterize a period when genetic information 'creates' a physical entity that differs in many respects from its predecessor, by reorganizing the chemical and physical structure of the material at its disposal, that is, the embryo and albumen in the egg stage, and the liquified caterpillar in the pupal stage. Thus a species, which may be viewed as a certain quantity of (genetic) information, requires certain nutrients, or combinations of elements, in order to give physical form to this information and add to it, in succeeding generations.

To clarify the point a bit – although the basic elements required for an organism to build itself, such as carbon, hydrogen and oxygen, are rather common, there is a realm of elements known as trace elements that, although abundant, are scattered almost in isolated molecular form over the Earth's surface. Even in this dispersed state, these elements can apparently exert a tremendous influence on their surroundings, so that they are indispensable components of an energy-flow system. For example, if a tenth or even a hundredth of 1 per cent of yttrium is added to cast iron, its hardness doubles, its resistance to wear quadruples and it becomes almost as strong as steel. Plants use such trace elements and extract them from the soil. However, trace element combinations depend more upon the area in which a plant grows, than its species. So plants of the same species from different areas do not necessarily possess an identical combination of trace elements. However, if a butterfly at two stages of its life dissolves itself in order to restructure the elements at hand, it follows that a butterfly probably depends more upon the chemical composition of a plant than the plant's species or other factors. It is probably for this reason that butterfly larvae from India at times refuse to feed off a plant species on which they have been bred, say in Thailand.

In fact, the variety of butterflies in an area varies in direct proportion to the diversity of local vegetation. One of the obvious reasons for this phenomenon is that, broadly speaking, individual butterfly species feed on different species of plant. Some plant species,

Whites and yellow, Pieridae *Three pairs of legs well developed, hindwings cover abdomen.*

Skippers, Hesperiidae *Hooked antennae originate far apart.*

Blues and Coppers, Lycaenidae *Small, generally with a tail at bottom of hindwing. Three pairs of legs well developed.*

Brush-footed butterflies, Nymphalidae *Both sexes with forelegs undeveloped.*

Punches and judies, Riodinidae *Female with three pairs of legs, male with forelegs undeveloped.*

Swallowtails, Papilionidae *Three pairs of legs well developed, hindwing cannot cover abdomen.*

such as the Tamarind and Himalayan Silver Oak, are host to several butterfly species, while others are avoided altogether.

Butterflies, being 'cold blooded', depend to a large extent on solar energy to fly, in that they reflect the sun's rays on to the body in order to achieve the internal body temperature required for flight (generally around 37° C, though higher in some hesperids).

Butterflies can survive excessive cold much better than they can survive heat. Himalayan butterflies, for example, regularly live through blizzards that may blow for days, with temperatures below zero. The first sun will be greeted by a ragged crew, emerging from crevices and crannies where it sheltered during the cold days. If the temperature, however, exceeds 45° C, most butterfly species find it difficult to survive, except those that have adapted to survive in such hot circumstances, such as the arabs (*Colotis* species).

Therefore, a change in thermal gradients in an area brought about by a reduction of vegetation can spell doom for butterfly species, even those not directly affected by the deforestation.

AREAS OF HIGH BUTTERFLY DIVERSITY

Of the 1,318 species found in India, roughly 340 occur on the peninsula, with most concentrated in the evergreen forests of the Western Ghats and the hills of southern India. The Gangetic plain has less than 100 species. Uttarakhand has around 450, and North-east India has well over 1,000.

There are butterfly gardens and parks throughout the country, such as Ovalekar Wadi in Thane, Maharashtra; the Butterfly Conservatory of Goa; Sammilan Shetty's Butterfly Park, Belvai, Karnataka; and Butterfly Safari Park, Thenmala, Kerala.

While a large number of national parks harbour a great diversity of butterflies, many do not permit visitors to travel on foot in them. Since it is almost impossible to identify or enjoy butterflies from a vehicle, there is no point visiting them for butterflies, unless it is confirmed beforehand that you are permitted to travel on foot, or at least to descend from vehicles. See Resources (p. 156) for information on knowledgeable people who can help with advice in this respect.

NOTE

The arrangement of families in the main section and the list of butterflies at the end differs; this is to facilitate reference in the field, while the list of species at the end follows the currently accepted arrangement of families. Several common bushbrowns and sailers, and many skippers have not been included, since they require handling or dissection to be properly identified.

Female upperside

Common Rose

Pachliopta aristolochiae 8–10cm

DESCRIPTION Sexes similar, but female usually has duller red markings than male and is often larger. Red thorax and abdomen on underside are distinctive. There are 3–5 white spots on hindwing, which vary individually in size. **DISTRIBUTION** Pakistan through India, Sri Lanka to Taiwan, the Philippines, Borneo and Java. **HABITATS AND HABITS** Common at low elevation but stragglers are found as high as 2,000m in western Himalaya. Avoids dense forests; common in scrubland and open areas. Distasteful to birds, since abdomen contains aristolochic acid. Males congregate at wet mud. Both sexes fond of flowers.

Male upperside

Male underside

Crimson Rose

Pachliopta hector 9–11cm

DESCRIPTION Sexes similar, but female usually has duller crimson spots on hindwing than male. There is never any white mark on hindwing. Red thorax and abdomen in combination with wing pattern are distinctive. Mimicked by *romulus* form of Common Mormon (see p. 27), which is easily distinguished by its black thorax and abdomen. **DISTRIBUTION** Endemic to India and Sri Lanka, as far north as West Bengal. Straggler on Andaman Islands and Uttarakhand. **HABITATS AND HABITS** Inhabits low-elevation forests and scrubland. Flight slow and fluttering. Often congregates in large numbers to roost. Distasteful to birds due to aristolochic acid in body. Both sexes fond of lantana flowers.

Male underside

Male upperside

Malabar Rose ■ *Pachliopta pandiyana* 10–13cm

DESCRIPTION Sexes similar. Both sexes larger than Common Rose (see opposite), and white patch on hindwing also larger than on Common. Mimicked by *stichius* form of Common Mormon (see p. 27), which is distinguished by its black abdomen. Very similar **Ceylon Rose** *P. jophon* (11–13cm; endemic to lowland evergreen forests in Sri Lanka, but has more extensive pale markings on both surfaces of forewing). **DISTRIBUTION** Endemic to seaward face of Western Ghats south of Goa. **HABITATS AND HABITS** Found in evergreen forests at low elevation. Forest insect, fluttering about within forests, with females searching for their low-growing food plants. Distasteful to birds due to aristolochic acid in abdomen. Both sexes fond of flowers.

Female underside

Female upperside

Ceylon Rose

Southern Birdwing ■ *Troides minos* 14–19cm

DESCRIPTION Male has plain yellow hindwings with broad black border. Female has row of large dark spots across hindwing. Singular in peninsular India. Females are the largest butterflies on the subcontinent. In **Sri Lankan Birdwing** *T. darsius* (14–17cm; Sri Lanka) both sexes have extensive black markings on hindwing. **DISTRIBUTION** Western Ghats from Maharashtra southwards. **HABITATS AND HABITS** Common at low elevation in humid forests. Visits flowers early in morning, and soars about treetops during daytime. Powerful flight occasionally takes it far from its habitat, and it can be seen in semi-urban gardens. Distasteful to birds.

Male upperside Sri Lankan

Male underside

Female underside

Female upperside Southern

Golden Birdwing ■ *Troides aeacus* 15–17cm

Female hindwing

DESCRIPTION Both sexes differ from Southern and Sri Lankan Birdwings (see p. 17) in having a narrower black border on hindwing. Differs from Common Birdwing (see below) and Southern Birdwing in having dark suffusion on yellow area of black marks along edge of upper hindwing in both sexes, which are lacking in Common (horizontal red arrow left). Female easily distinguished by dark base of space 1b on hindwing (vertical red arrow), lacking in all other similar species. **DISTRIBUTION** Himalaya, at 1,200–2,750m in west Himalaya, descending to plains in Assam and Myanmar; to Taiwan and Malaysia. **HABITATS AND HABITS** Fond of flowers of lantana, thistles and horse chestnut. Males can be seen settled on damp grass early in the morning. In North-east India and Myanmar, attracted to damp sand. Distasteful to birds.

Male upperside

Female upperside

Male underside

Common Birdwing ■ *Troides helena* 13–17cm

DESCRIPTION Both sexes without dark suffusion on upper hindwing. Male usually has a spot at bottom of under hindwing (red arrow). In female, base of space 1b on hindwing is yellow. **DISTRIBUTION** Nepal through North-east India to Myanmar, southwards to Malaysia. **HABITATS AND HABITS** Common at low elevation in open country and forests. Both sexes fond of flowers of lantana. Males visit wet sand. Often found in company of Golden Birdwing (see above). Distasteful to birds.

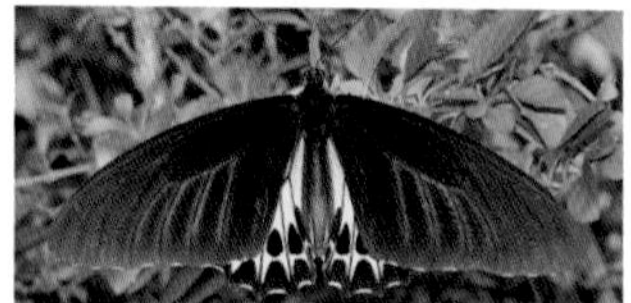

Female upperside form eumagos

Male underside

Female underside

Male upperside

Lesser Batwing ■ *Atrophaneura aidoneus* 10–12cm

Male underside

DESCRIPTION Sides of head and body pinkish, not red as in Common Batwing (see below). Lesser male has white scent-fold edged with pink along inner edge of hindwing, which is usually folded over; this is black in Common male. Lesser female has plain dark forewing and hindwing, unlike Common female, which has pale area on lower half of forewing; both sexes of Lesser have uniform dark shade on hindwing. **DISTRIBUTION** West Himalaya from Uttarakhand to Vietnam. **HABITATS AND HABITS** Forest insect, occurring from low elevation to 2,500m. Females usually seen fluttering about inside forest. Occurs in company of windmills and birdwings, but not found in open country. Males visit wet sand. Both sexes fond of flowers of thistle and horse chestnut.

Female upperside

Male showing scent-folds

Common Batwing ■ *Atrophaneura varuna* 9–13cm

DESCRIPTION Sides of head red. Underside of body red, not pinkish as in Lesser Batwing (see above). Scent-fold along inner edge of hindwing black and fringed with hair in male. Female has pale area at bottom of upperside forewing. Hindwing cell darker than rest of wing in both sexes. Subspecies *zaleucus* (Southern Myanmar) has broad white area on hindwing. In old female, body fades to pinkish, and darker inner half of hindwings is then useful in distinguishing it from Lesser female. **DISTRIBUTION** Uttarakhand to North-east India, Vietnam and Malaysia. **HABITATS AND HABITS** Usually at low elevation. Forest insect, rarely found in open country. Distasteful to birds due to aristolochic acid stored in body. Males visit wet sand. Both sexes fond of flowers of lantana.

Male underside

Female upperside

Male upperside

Male underside zaleucus

Common Windmill ■ *Byasa polyeuctes* 11–14cm

DESCRIPTION Sexes similar. Hindwing with one large, angular quadrate white spot, with or without white streak below it. There may be more pale spots on hindwing, especially in females. Similar **Great Windmill** *B. dasarada* (10–14cm; same distribution as Common) has the large white spot on hindwing in form of a capital 'D', with an extra prominent white spot above it in west Himalaya populations; spot usually lacking in east Himalaya and Myanmar populations. **DISTRIBUTION** Murree in Pakistan to Taiwan, Vietnam and Thailand. **HABITATS AND HABITS** Forest insect that ventures into scrubland in search of flowers. Flight slow, fluttering within forest, soaring around treetops during day. Distasteful to birds, containing aristolochic acid in body. In North-east India and Myanmar, males visit wet sand. Both sexes fond of flowers of clematis, lantana, horse chestnut and thistles.

Male underside

Male upperside

Male underside Great

Tawny Mime ■ *Papilio agestor* 8.3–12cm

DESCRIPTION Sexes similar. Bright chestnut area on hindwing and white-spotted black abdomen characterize this species. Subspecies *govindra* is greyer, with darker chestnut area on hindwing and complete row of white spots on it. These spots may be entirely absent in some individuals of east Himalaya subspecies *agestor*. Similar **Lesser Mime** *P. epicydes* (7–9cm; Nepal to Taiwan), which lacks chestnut area on hindwing, easily distinguished by prominent yellow spot at bottom of hindwing on both surfaces. Occurs in spring at up to 2,000m. **DISTRIBUTION** Occurs along Himalaya from Jammu and Kashmir to North-east India, on to southern China, Taiwan and Malaysia. **HABITATS AND HABITS** Inhabits forest from low elevation to 2,400m. Single annual brood; on the wing in spring. Exactly mimics Chestnut Tiger (see p. 123) in appearance and behaviour. Males often territorial on hilltops, and attracted to wet sand. Both sexes fond of jasmine, thistle and horse chestnut flowers.

Male upperside agestor

Male upperside govindra

Male underside

Lesser Mime

Blue-striped Mime ■ *Papilio slateri* 8–10cm

DESCRIPTION Sexes similar. Yellow spot at bottom of hindwing on both surfaces immediately distinguish this species from the blue crows (*Euploea* spp.) that it mimics. **DISTRIBUTION** Sikkim to China and southwards to Borneo. **HABITATS AND HABITS** Single, spring brood at low elevation. Females rarely found, usually confining themselves within forests. In flight, both sexes mimic members of the blue crows remarkably well. Males attracted to wet sand. Both sexes fond of flowers.

Male upperside

Male underside

Common Mime ■ *Papilio clytia* 9–10cm

DESCRIPTION Sexes similar. Four subspecies and seven forms: for subspecies *clytia*, brown forms *clytia*, *panope*, *janus*, *papone* and *commixtus* resemble crows (*Euploea*); white form *dissimilis* and North-east Indian, darker form *dissimillima* resemble Blue Tiger (see p. 123). All forms have distinctive small yellow spot at bottom of upperside hindwing and row of yellow spots on underside hindwing. Sri Lankan subspecies *lankeswara* has reduced white spots on forewing; its white form *dissimila* resembles *dissimilis*; Andaman Islands subspecies *flavolimbatus* has yellow spots on upperside. **DISTRIBUTION** Throughout India, Myanmar and Sri Lanka, except arid parts to Timor. **HABITATS AND HABITS** Ascends to 2,750m in Himalaya. Flight powerful at times, especially when males spend time hill-topping; at other times, flight exactly resembles its tiger and crow models, but it can be distinguished by its larger size. Males visit wet mud. Both sexes fond of flowers.

Underside clytia

Upperside clytia

Underside dissimilis

Upperside dissimilis

Blue Mormon ■ *Papilio polymnestor* 12–15cm

DESCRIPTION Sexes similar. Blue on upperside makes this a singular butterfly on the Indian peninsula and in Sri Lanka. In North-east India, *polymnestroides* form of Great Mormon (see below) occurs with this species. Females often have red patch at base of upperside forewing. **DISTRIBUTION** Endemic to Nepal, India and Sri Lanka. **HABITATS AND HABITS** Common at low elevation in gardens and forests. Flies low and fast along forest edges and paths, stopping briefly to visit lantana and other flowers. Males come to wet sand, and in Kerala can be found mud-puddling even at dusk in May. Possibly hybridizes in the wild with Great Mormon in North-east India.

Male upperside

Male underside

Great Mormon ■ *Papilio memnon* 12–15cm

DESCRIPTION One male and three female forms. In all forms base of wings is red on underside. Female form *alcanor* has hindwing tails and yellow abdomen; female form *agenor* similar to *alcanor*, but lacks tails and has black abdomen; female form *butlerianus* similar to male, but has pale lower third of forewing; form *memnon* tailed or tailless, with white hindwing and row of black spots along edge; form *polymnestorides* (male and female) resembles Blue Mormon (see above). **DISTRIBUTION** Nepal to Japan, Indonesia, the Flores and the Philippines. **HABITATS AND HABITS** Forest insect, found at low elevation, where it may be seen flying swiftly and low along forest paths and hedgerows, and in gardens. Females more often found within forests, searching among low bushes for a suitable plant on which to lay their eggs. Among the most variable species known, with numerous female forms over its range. Males often visit wet sand. Both sexes fond of flowers.

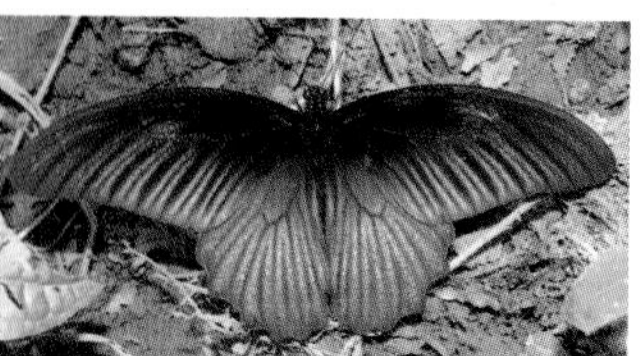

Male upperside

Male underside

Female butlerianus

Female alcanor

Female underside alcanor

Spangle ■ *Papilio protenor* 10–14cm

Female underside

DESCRIPTION Body black, and no red at base of wings. Female lacks pale grey bar along costa of upper hindwing. **Redbreast** *P. alcmenor* (11–13cm; Uttarakhand to Hainan) male has red base of underside hindwing, and red and grey lunule on upperside hindwing; **Yellow-crested Spangle** *P. elephenor* (11–13cm; Assam) has yellow head and prominent yellow stripe along side of abdomen. **DISTRIBUTION** Along Himalaya from Kashmir to North-east India, Indo-China, Korea and Japan. **HABITATS AND HABITS** Ascends to 2,750m. Flies swiftly along forest paths, rarely crossing open spaces. Distasteful to birds. Males attracted to wet mud. Both sexes fond of flowers.

Male upperside

Yellow-crested

Underside Redbreast

Upperside Redbreast

Common Peacock ■ *Papilio bianor* 9–13cm

DESCRIPTION Upper hindwing blue patch may or may not be connected to inner margin, but never by sharply defined, thin green line (red arrow below). Female lacks two woolly black streaks across lower half of diffuse green band on upper forewing (red circle below). In western Himalaya, spring brood is small in size. On underside, inner half of forewing is plain, with no markings in cell. **Krishna Peacock** *P. krishna* (12–13cm; Nepal to China and Myanmar) has prominent green band across forewing on upperside, represented as similar yellow band on underside of forewing. **DISTRIBUTION** Pakistan along Himalaya to Korea, Japan and Thailand. **HABITATS AND HABITS** Ascends to 1,600m. Swarms at wet mud after monsoon; visits lantana and marigold flowers. Voted the most beautiful Indian butterfly. Distasteful to birds.

Male underside

Male upperside Krishna

Female upperside

Male upperside

Paris Peacock ■ *Papilio paris* 9–14cm

Male tamilana

DESCRIPTION Sexes similar. Green band on upper forewing narrow and sharply defined; almost absent in some southern Indian individuals. Large blue patch on upper hindwing connected to inner margin by narrow, sharply defined green band. On underside, inner half of forewing is plain, with no markings in cell. Southern Indian subspecies *tamilana* rather larger than Himalayan populations. **DISTRIBUTION** Western Ghats south of Maharashtra; Himalaya from Uttarakhand eastwards to South-east Asia. **HABITATS AND HABITS** Ascends to 1,500m in Himalaya, to top of south Indian hills. Flight powerful, about the treetops. Generally, flight more ragged than that of Common Peacock (see p. 23). Males visit wet sand. Both sexes fond of flowers.

Male upperside

Male underside

Common Banded Peacock ■ *Papilio crino* 8–10cm

Underside Malabar Banded

DESCRIPTION Sexes similar. Narrow, sharply defined peacock-green bands across upperside of both wings. Some males have woolly streaks across green band on forewing. **Malabar Banded Peacock** *P. buddha* (9–10cm; endemic to western face of Western Ghats from Goa southwards) has broader green bands and orange mark along costa of upper hindwing. **DISTRIBUTION** Sri Lanka; in open forests across drier parts of peninsular India south of Orissa and Karnataka. **HABITATS AND HABITS** A butterfly of hot, dry regions, never found on seaward face of Western Ghats. Flight swift, near the ground. Not known to be attracted to wet sand. Both sexes settle frequently on flowers.

Upperside Malabar Banded

Male underside

Male upperside

Malabar Raven ■ *Papilio dravidarum* 8–10cm

DESCRIPTION Similar to form *clytia* of Common Mime (see p. 21), but lacks all yellow marks. Distinguished from Common Crow (see p. 124) by arrowhead-shaped white marks on hindwing. Some individuals have white spot in middle of forewing cell. **DISTRIBUTION** Evergreen forest along seaward face of Western Ghats from Goa southwards to Kerala. **HABITATS AND HABITS** Ascends to shola forest in Anamalai Hills. Males attracted to wet sand, where they may be seen even at dusk in May. Both sexes fond of flowers.

Male upperside

Male underside

Red Helen ■ *Papilio helenus* 11–13cm

DESCRIPTION Sexes similar. On underside, row of red spots with large white patch on hindwing is distinctive. Red spots may be present or entirely absent on upper hindwing, though females usually have row of red spots. **DISTRIBUTION** Sri Lanka; Western Ghats south of Gujarat; Himalaya from Uttarakhand eastwards to North-east India and South-east Asia. **HABITATS AND HABITS** Common at low elevation, where it inhabits forests in regions of heavy rainfall. Swift flight, often about treetops, but males readily descend to wet sand. Both sexes fond of flowers.

Female upperside

Male underside

Yellow Helen ■ *Papilio nephelus* 11.5–13cm

DESCRIPTION Sexes similar. Under hindwing has row of yellow spots; upper hindwing has no markings except for central white patch. **DISTRIBUTION** Nepal and Odisha to South-east Asia. **HABITATS AND HABITS** Common in dense forest at low elevation. Powerful flight, about the level of treetops and tall bushes. Often seen flying along streams and river edges. Males attracted to wet sand. Both sexes fond of flowers.

Underside

Upperside

Underside

Malabar Banded Swallowtail ■ *Papilio liomedon* 9–10cm

DESCRIPTION Sexes similar. Singular in peninsular India. Nearly identical **Burmese Banded Swallowtail** *P. demolion* (9–10cm; Myanmar south to Borneo and the Philippines). **DISTRIBUTION** Endemic to seaward face of Western Ghats south of Karnataka. **HABITATS AND HABITS** Found in hills, usually confined within forest, where it is content to fly in shade of trees and settle on prominent leaves. Visits flowers, but does not seem to have been recorded at wet mud.

Upperside

Male underside

Common Mormon ■ *Papilio polytes* 9–10cm

DESCRIPTION One male and three female forms. Form *cyrus* looks like male (both sometimes with red spots on upperside hindwing); *stichius* resembles Common Rose (see p. 16); *romulus* resembles Crimson Rose (see p. 16). All forms distinguished from their models by black abdomen (blue circle right). Males sometimes have yellow marks on hindwing instead of red ones. *Romulus* form of female restricted to Indian subcontinent. **DISTRIBUTION** Pakistan to Sri Lanka throughout India, eastwards to the Philippines and Indonesia. **HABITATS AND HABITS** Ascends to 1,600m in Himalaya. Flight rapid, near the ground. Mimetic females at times perfectly imitate flight of their models. Distasteful to birds. Males often congregate by the hundred on wet sand. Both sexes fond of flowers.

Underside romulus

Underside stichius

Male underside

Female upperside

Male upperside

Lime ■ *Papilio demoleus* 8–10cm

DESCRIPTION Sexes similar. Singular species in India. Yellow ground colour darkens with age. Its closest relatives are in Africa and west Asia. **DISTRIBUTION** Throughout Pakistan, India, Myanmar and Sri Lanka to Australia. **HABITATS AND HABITS** Ascends to 1,600m in Himalaya. Flight swift, near the ground. Males congregate by the hundred on damp mud after south-west monsoon. Both sexes fond of flowers.

Male underside

Male upperside

Common Yellow Swallowtail ■ *Papilio machaon* 7.5–9cm

DESCRIPTION Distinguished from similar **Chinese Yellow Swallowtail** *P. xuthus* (7.5–9cm; Arunachal Pradesh to Japan) by plain dark forewing cell, which has yellow lines in *xuthus*. **Southern Swallowtail** *P. alexanor* (7.5–9 cm; Baluchistan to Europe) has black marks across forewing. Above 4,500m in Ladakh and Sikkim, Common Yellow Swallowtail lacks tails on hindwings. **DISTRIBUTION** Throughout Himalaya above 1,200m; Europe to Siberia. **HABITATS AND HABITS** Flies low over meadows; never found in dense forests. Males patrol patches of meadow, especially along ridges, where they await passing females. They also visit wet sand. Both sexes fond of flowers.

Underside

Male upperside

Southern Swallowtail upperside

Chinese Yellow Swallowtail upperside

Spot Swordtail ■ *Graphium nomius* 7.5–9cm

DESCRIPTION Sexes similar. Large size and long tails on hindwing are distinctive. Row of pale spots along edge of forewing give species its common name. Very similar **Chain Swordtail** *G. aristeus* (7–8cm; Sikkim to Australia) is identical, but distinguished by pale band on forewing, which is in form of chain (red arrow). **DISTRIBUTION** Dry deciduous forests on eastern side of Western Ghats and Deccan plateau; low elevations along Himalaya from Himachal Pradesh eastwards to Myanmar and southwards to Sri Lanka; to China and Indo-China. **HABITATS AND HABITS** Males visit damp mud, where hundreds may gather after monsoon. Once recorded migrating westwards in large numbers in Rajasthan. Both sexes visit flowers.

Chain Swordtail

Male upperside

Male upperside

Fivebar Swordtail ■ *Graphium antiphates* 8–9.5cm

DESCRIPTION Sexes similar. Large size, long hindwing tails, green underside and white upperside distinctive. Five black bars along upper edge of forewing, which give it its common name. Nearly identical **Andaman Swallowtail** G. *epaminondas* (8–9.5cm) endemic to Andaman Islands. **DISTRIBUTION** Sri Lanka; Western Ghats south of Goa; Himalaya east of Nepal to South-east Asia, except Andaman and Nicobar Islands. **HABITATS AND HABITS** Restricted to low elevation. Flight very swift, often rather high, when it looks more like a pierid than a swallowtail. Males attracted to wet mud. Both sexes fond of flowers.

Male underside

Male upperside

Sixbar Swordtail ■ *Graphium eurous* 6–7cm

DESCRIPTION Sexes similar. Upper forewing has six black bars along costa. On under hindwing, chain of black-bordered yellow spots across middle. This is reduced to two spots in the form of a spectacle in **Spectacle Swordtail** G. *mandarinus* (6.5–7.5cm; occurs with Sixbar over most of it range, except Jammu and Kashmir). The two species are indistinguishable in terms of behaviour and habits. **DISTRIBUTION** Jammu and Kashmir (Dras) to Thailand. **HABITATS AND HABITS** Dense broadleaved forests at 1,200–3,000m in west Himalaya; at lower elevation in eastern Himalaya. In the morning hours, flight is swift around the canopy, but by midday, butterflies descend to mid-level, then behave very like cabbage whites (*Pieris* spp.), with which they are easily confused. Males come frequently to wet sand on riverbanks. Both sexes fond of flowers of *Viburnum cotinifolium*, *Eupatorum*, *Buddleia*, *Anaphalis* and similar plants.

Upperside

Underside

Underside Spectacle

Glassy Bluebottle ■ *Graphium cloanthus* 8.5–9.5cm

DESCRIPTION Sexes similar. Pale blue areas translucent. Pattern on wings is singular on Indian peninsula. Similar **Fourbar Swordtail** G. *agetes* (7.5–9cm; Sikkim to China and Borneo) has similarly translucent wings but a different pattern. **DISTRIBUTION** Pakistan east along Himalaya to North-east India and Myanmar, China and Taiwan, at 400–2,750m. **HABITATS AND HABITS** Flight jerky, swift and generally around tops of trees. Males spend hours patrolling favourite hilltops or trees. They come to wet sand, where several may gather, though large congregations of the species are unknown. Both sexes fond of flowers of thistle, horse chestnut and buddleia.

Male upperside

Male underside

Underside and upperside Fourbar

Common Bluebottle ■ *Graphium sarpedon* 8–9cm

DESCRIPTION Sexes similar. On forewing, black border is unmarked. **Southern Bluebottle** G. *teredon* (8–9cm; Gujarat to Sri Lanka along seaward face of Western Ghats) has tooth-like short tail at bottom of hindwing more developed than in Common. **DISTRIBUTION** Along Himalaya from Pakistan to Myanmar, north to Japan and southwards to Australia. Not recorded from Gangetic plain. **HABITATS AND HABITS** In its range, common in humid forest from low elevation to 2,750m. Flight rapid, jerky and generally around tops of trees. Males can spend hours circling a favourite tree on an exposed hilltop, waiting for passing females. Both sexes come readily to flowers. Males visit wet mud.

Common

Male upperside Southern

Male underside Common

Common Jay ■ *Graphium doson* 7–8cm

DESCRIPTION Sexes similar. On forewing, black border has row of blue spots. Costal black bar on underside hindwing does not connect with basal black band (vertical red arrow). **Scarce Jay** *G. albociliatis* (Assam to Myanmar and Laos), and **Great Jay** *G. eurypylus* (7.5–9cm; North-east India to South-east Asia) always have costal bar connected to basal band. **DISTRIBUTION** Most of India, Myanmar and Sri Lanka. Has recently colonized Delhi and adjoining parts of Gangetic plain. **HABITATS AND HABITS** In its range, common in forested areas of moderate to heavy rainfall, and confined to low elevation. Flight rapid, jerky, and very similar to flights of Common and Southern Bluebottles (see opposite), with which it can be confused when flying. Males visit wet mud. Both sexes fond of flowers.

Male underside

Male underside

Male upperside

Male Scarce

Veined Jay ■ *Graphium chirnoides* 7.5–10cm

DESCRIPTION Sexes similar. Underside has chrome-yellow spots. Upperside forewing has broader pale marks compared to Common Jay's. **DISTRIBUTION** Nepal eastwards along Himalaya to North-east India, Myanmar and Malaysia. **HABITATS AND HABITS** Confined to low elevation. Flight jerky, rapid, and usually around tops of trees. Males attracted to wet mud, and often found as part of large congregations of other butterfly species. Both sexes fond of flowers.

Male upperside

Male underside

Tailed Jay ■ *Graphium agamemnon* 8.5–10cm

DESCRIPTION Sexes similar. Hindwing has tail. Bright green spots on wings. Very similar **Spotted Jay** *G. arycles* (7–8cm; Manipur to Borneo and the Philippines) lacks hindwing tails. **DISTRIBUTION** All over India and Sri Lanka to Sulawesi and Australia. **HABITATS AND HABITS** In its range, occurs in humid forests and rural landscapes. Common at low elevation. Flight swift and skipping, and butterfly rarely settles. Females ascend Himalaya to more than 2,000m in search of host plants. Both sexes fond of flowers.

Male underside

Spotted

Upperside

Green Dragontail ■ *Lamproptera meges* 4–5.5cm

DESCRIPTION Sexes similar. Cannot be confused with any other butterfly, except **White Dragontail** *L. curius* (4–5 cm; Assam to Indonesia and Palawan, ascending to 1,200m in eastern Himalaya), which, as the name implies, has a white rather than green band across wings. In Green, broad black band across middle of forewing is of even width throughout its length, while band narrows in White. **DISTRIBUTION** Assam to the Philippines and Indonesia. **HABITATS AND HABITS** Occurs at low elevation, usually near water. Flight rapid and skipping, so that butterfly somewhat resembles a dragonfly. Often seen near water, where males settle readily on wet sand. Both sexes settle on low vegetation to bask, and visit flowers occasionally.

White

Male upperside

Male underside

Yellow Gorgon ■ *Meandrusa payeni* 11–13cm

Male underside

DESCRIPTION Sexes similar. Shape, colour and pattern of wings are singular. Related **Brown Gorgon** M. *lachinus* (10.5–11.5cm; Uttarakhand along Himalaya to Myanmar, Thailand and southern China, above 1,800m in western Himalaya, at lower elevation in east) is also a singular butterfly. **DISTRIBUTION** Sikkim eastwards to Hainan and south to Borneo and Java. **HABITATS AND HABITS** Inhabits dense broadleaved forests in regions of heavy rainfall. Little known of its habits, except that it likes to settle on wet mud, preferably where there are dry leaves on the ground so that it merges with them. Flight swift, generally high up among the treetops. Both sexes attracted to flowers.

Male upperside

Brown Gorgon male

Kaiser-i-Hind ■ *Teinopalpus imperialis* 9–12cm

DESCRIPTION Males small, with yellow patch and single tail to hindwing. Females larger with two prominent tails, and lack yellow patch on hindwing. **DISTRIBUTION** Nepal to northern Myanmar, China and Vietnam. **HABITATS AND HABITS** Montane butterfly, usually found about 2,000m above sea level. Males fond of hill-topping, spending the morning hours circling a prominent tree and sparring with passing butterflies. Females less often seen and seem to prefer forests. Both sexes attracted to over-ripe fruits.

Female upperside

Male upperside

Male underside

Bhutan Glory ■ *Bhutanitis lidderdalii* 9–11cm

DESCRIPTION Sexes similar. Singular species that can only be confused with **Ludlow's Bhutan Glory** *B. ludlowi* (Bhutan and eastern Arunachal Pradesh), from which it is easily distinguished by presence of four prominent tails on hindwing, while *ludlowi* has only three tails. **DISTRIBUTION** Sikkim to northern Thailand and southern China. **HABITATS AND HABITS** Restricted to elevation above 1,800m in Himalaya, where its larval host plant grows. Found in dense forest, where it sails about above the treetops with a weak, fluttering flight. Females look remarkably like the windmills (*Byasa* spp.) at times, especially when laying eggs. Both sexes rarely visit flowers.

Upperside

Ludlow's

Common Red Apollo ■ *Parnassius epaphus* 4–6cm

DESCRIPTION Sexes similar. Identical to **Keeled Apollo** *P. jacquemontii* (5.5–7.5cm; Afghanistan to Uzbekistan and Kyrgyzstan) except for antennae, which are prominently chequered black and white in Common Red and less so in Keeled. Red spots on hindwing never white centered in Common Red, and often so in Keeled. Himalayan populations of **Scarce Red Apollo** *P. actius* (6cm; Jammu and Kashmir to Afghanistan and Kazakhstan) distinguished from Common Red only by structure of hindwing veins (red arrow), which are separate in this species and joined in Common Red. **DISTRIBUTION** Pakistan to China and eastern Himalaya. **HABITATS AND HABITS** Found above 3,800m. Inhabits scree slopes and barren places, where it spends the night under small, flat stones. Flight fluttering, low but surprisingly agile. Females often mated soon after emerging from their pupae, and before they have expanded their wings. Both sexes fond of flowers, and may be found on small patches of meadow, where they settle frequently in their wanderings about slopes. Occasionally, males are attracted to water seepages. Mated females have pouch at tip of abdomen, to prevent promiscuity. Shape of pouch is often the only means of identifying species in this group.

Scarce Red

Male upperside Keeled

Common Red

Scarce Red

Common Blue Apollo ■ *Parnassius hardwickii* 5–6.5cm

Male

DESCRIPTION Sexes similar, but females often more heavily marked than males. Easily distinguished from all other members of the genus by row of black-bordered blue spots along edge of hindwing. Some individuals lack almost all their markings. Summer brood usually brightly coloured, while spring and autumn broods are less brightly marked. **DISTRIBUTION** Confined to Himalaya. The most common species of the genus and the only member found on southern face of Himalaya in monsoon zone. Has been recorded at Mashobra near Shimla, Himachal Pradesh. **HABITATS AND HABITS** Occurs above 2,500m. The only member of the genus that is known to overwinter as adults in forests below treeline, where it can often be seen on sunny winter days. Males often sport in updrafts on cliffs, rising to top on wind currents and descending to repeat the performance. Flight weak and fluttering, but capable of making rapid progress. Fond of visiting flowers.

Female upperside

Male

Regal Apollo ■ *Parnassius charltonius* 8–9cm

DESCRIPTION Sexes similar. Large size of butterfly and large red spot with blue-crowned black spots along border of hindwing immediately distinguish this species. Smaller **Greater Banded Apollo** *P. stenosemus* (6.4–6.8cm; Ladakh to Pakistan above 4,000m) has smaller red spot on hindwing that is never white centred (horizontal green arrow), and black spots along edge of hindwing are usually white centred. **Lesser Banded Apollo** *P. stoliczkanus* (5.2–6 cm; Pakistan to Nepal above 4,000m) distinguished by smaller size than that of both the preceding species, and darker suffusion along border of hindwing than in either of them. **DISTRIBUTION** Nepal to China, Pakistan and Uzbekistan. **HABITATS AND HABITS** Occurs above 4,000m. Flight powerful across hillsides, but species is fond of flowers and settles often to bask. Gathers in caves and under overhangs in bad weather.

Lesser Banded

Regal

Greater Banded

Common Emigrant ■ *Catopsilia pomona* 5.5–8cm

Male upperside

DESCRIPTION Sexes dissimilar. Three forms, all without mottling on underside; male and female *crocale* have plain, unmarked underside; male and female *pomona* have red-ringed silver spots in centre of wings on underside; female form *catilla* has large purple blotches on underside of hindwings. Male upperside chalky-white with narrow black border to forewing; female *crocale* upperside white with broad black markings on forewing that are very variable in extent; female *pomona* and *catilla* uppersides identical, lemon-yellow with dark border and dark spot on forewing. Similar **Orange Emigrant** C. *scylla* (6–6.5cm; Sri Lanka, Myanmar to the Philippines and Sulawesi) has white forewings and orange hindwings on upperside. On underside, forewing apex is unmarked. **DISTRIBUTION** Throughout Indian subcontinent to Australia. **HABITATS AND HABITS** Flight powerful, skipping and generally at level of treetops, but descends readily to flowers and wet sand. Swarms during rainy season.

Female crocale *Female* pomona

Male underside crocale

Male underside pomona

Female underside catilla

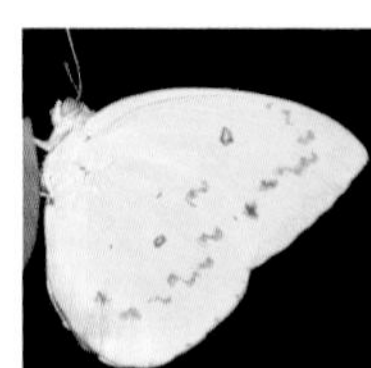

Orange

Female underside crocale

Female underside pomona

Orange

Mottled Emigrant ■ *Catopsilia pyranthe* 5–7cm

DESCRIPTION Sexes dissimilar. Underside pale greenish, mottled with dark lines. Form *pyranthe* has small dark speck in centre of wings on underside; form *gnoma* has red ring. Form *pyranthe* has black antennae, form *gnoma* red antennae. Form *gnoma* has upperside black forewing border in form of separate dark marks. In addition to all the markings of male, female has indistinct series of 3–4 dark specks below apex of upperside forewing. **DISTRIBUTION** Throughout Indian subcontinent to Australia. **HABITATS AND HABITS** Flight similar to Common Emigrant's (see opposite), but Mottled can be easily distinguished by smaller size and greenish-white shade of butterfly, which is yellowish-white in Common. Both sexes visit flowers and are fond of wet sand, where males sometimes congregate in huge numbers.

Female gnoma

Male pyranthe

Female gnoma

Male pyranthe

Female pyranthe

Himalayan Brimstone ■ *Gonepteryx nepalensis* 6–7cm

DESCRIPTION Male's upperside butter-yellow, female's creamy-white. Pointed forewing apex distinctive for the genus. **DISTRIBUTION** Along Himalaya from Pakistan (Chitral) to Myanmar. **HABITATS AND HABITS** Occurs at 1,000–2,750m. Fond of flowers. Rarely, males visit wet sand. Among the longest lived butterflies, existing up to ten months in adult stage, much of which is spent hibernating. Males of British member of this genus were probably originally referred to as 'butter (-coloured) flies' in English villages, where it was considered good luck if the first butterfly you saw in spring was yellow. With wings closed, butterfly resembles a dead leaf. This resemblance is exploited when avoiding predators or, in the case of females, pestering males. A butterfly simply drops off its perch and wafts down without opening its wings, very like a dead leaf.

Female underside

Female upperside

Male upperside

Tree Yellow ■ *Gandaca harina* 3.5–4.5cm

DESCRIPTION Male pale yellow, female creamy-white. Wings rounded. Underside unmarked; upperside with narrow black border to forewing. Larger than most grass yellows. **DISTRIBUTION** Central Nepal through North-east India to the Philippines and Borneo. **HABITATS AND HABITS** Occurs at low elevation. Restricted to areas of heavy rainfall. Jerky flight. Females usually found within forest, where they fly a few metres above the ground. Males fond of settling on wet sand. Both sexes visit flowers of lantana.

Male underside

Female underside

Female upperside

One-spot Grass Yellow ■ *Eurema andersoni* 3.8–4.5cm

DESCRIPTION Only one dark spot in cell of underside forewing (red circle). In peninsular Indian subspecies *shimai*, upperside hindwing dark border is very narrow, while it is broad in Himalayan and North-east Indian subspecies *jordani*. Dark mark on costa of underside hindwing is directed towards spot at end of cell (red circle), which distinguishes this from individuals of Common Grass Yellow (see opposite) with one cell-spot. Similar **Sri Lankan Grass Yellow** *E. ormistoni* (4–4.5cm; Sri Lanka) distinguished by costal mark on underside hindwing being obscure. On upperside hindwing, dark border is thread-like in female, expanding to short, dark wedges on veins in male. **DISTRIBUTION** Western Ghats from Karnataka southwards; Himachal Pradesh eastwards to North-east India, Myanmar eastwards to Taiwan and southwards to Indonesia and Borneo. **HABITATS AND HABITS** Occurs at low elevation, ascending Himalaya to 1,400m. Resident of humid evergreen forests and in South India often found about the canopy, unlike other members of the genus. Males attracted to wet sand. Both sexes fond of flowers.

Underside jordani

Underside shimai

Sri Lankan

Three-spot Grass Yellow ■ *Eurema blanda* 4–5cm

DESCRIPTION Sexes similar. Female has broader black marginal bands on upperside than male. Always has three black marks in cell of underside forewing (red circle). Dry-season form has rust-coloured patch below underside forewing apex. Similar to Common Grass Yellow (see below) with equally variable black margin on upperside forewing. **DISTRIBUTION** Sri Lanka, Gujarat to Kerala, Uttarakhand to North-east India, Andaman and Nicobars, to Japan, the Philippines and Papua New Guinea. **HABITATS AND HABITS** Often found singly within forest, where flight is strong and purposeful a few metres above the ground. Fond of wet sand and flowers.

Underside wet-season form

Underside dry-season form

Upperside

Common Grass Yellow ■ *Eurema hecabe* 4–5cm

DESCRIPTION Sexes similar. Female often has broader black margins to wings than male. Very variable. Dry-season form often has rust-coloured patch near apex of underside forewing. Underside forewing has two dark marks in cell (red circle), one or both of which may be absent. Black border on upper forewing variable, with or without an excavation in middle. **DISTRIBUTION** Common all over Pakistan, India and Sri Lanka, westwards to Africa and eastwards to Australia. **HABITATS AND HABITS** Favours open areas, paths, stream edges and hedgerows, where it flies about rather near the ground. Flight usually weak, but can be purposeful above level of canopy at times. Fond of wet sand, where hundreds of males congregate during wet season. Both sexes fond of flowers.

Underside wet-season form

Male upperside

Male upperside

Underside dry-season form

Small Grass Yellow ■ *Eurema brigitta* 3–4cm

DESCRIPTION Female has broader black borders to upperside wing margins than male. Upperside forewing black margins always have evenly curved inner edge (upper red arrow) and reach tornus (lower red arrow). Two tiny black specks at end of cell on underside forewing. Some individuals have pink edging to wings. Black margins broader in wet-season form. **DISTRIBUTION** Africa through Indian subcontinent and Sri Lanka to Australia. **HABITATS AND HABITS** Ascends to almost 4,000m in Himalaya. Among the few butterflies active on misty days. Flight weak and ragged, yet capable of covering considerable distances. Fond of wet sand. Both sexes visit flowers.

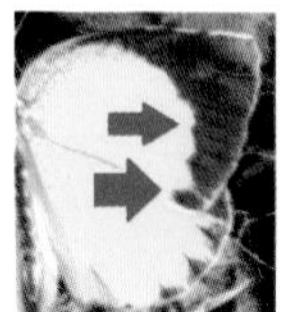

Male wet season

Male underside

Female dry season

Underside with pink border

Spotless Grass Yellow ■ *Eurema laeta* 3–4.5cm

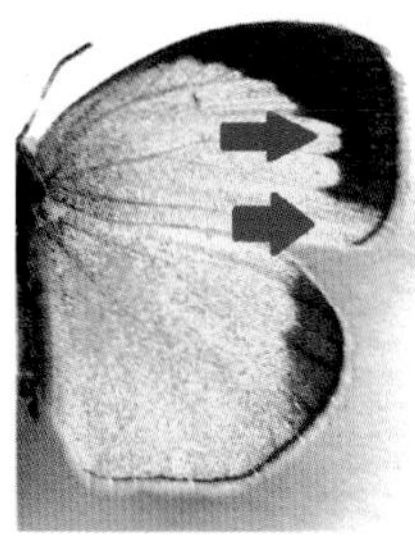

Female venata

DESCRIPTION Sexes similar. Female has broader dark margins in wet-season form. Species lacks most of black markings on underside. Dry-season form is singular, with sharp tip to forewing apex. Wet-season form very like Small Grass Yellow, (see above) but black border on upperside forewing does not reach tornus (red arrows). **DISTRIBUTION** Throughout Pakistan, India and Sri Lanka to Japan and Australia. **HABITATS AND HABITS** Ascends Himalaya to more than 2,000m. Males not usually gregarious like those of Common Grass Yellows (see p. 39) and Small Grass Yellows. Flight more powerful than Small Grass Yellow's. Males attracted to damp sand. Both sexes fond of flowers.

Male venata

Male laeta

Underside laeta

Underside venata

Dark Clouded Yellow ■ *Colias fieldii* 4.5–6.5cm

DESCRIPTION Upperside ground colour orange. Male has plain black border to upper forewing; female has border spotted with yellow. Very rarely, upperside ground colour is yellow, not orange. **Fiery Clouded Yellow** C. *eogene* (4–5.5cm; meadows at high elevation of 3,500m and above in inner ranges of western Himalaya) darker orange than Dark Clouded. **DISTRIBUTION** From Iran to China and Ussuri, throughout Himalaya, and descending to plains in Delhi and Assam in winter. **HABITATS AND HABITS** Ascends to more than 4,000m in Himalaya. Always avoids shady forests. Frequently seen in spring on meadows, in fields and along sunny forest paths. Flight rapid, keeping low over grass, and never ascending to canopy level of a forest. Both sexes fond of flowers.

Male upperside

Female underside

Female upperside

Pale Clouded Yellow ■ *Colias erate* 4.5–5.5cm

DESCRIPTION Variable. Male's upperside black forewing border has yellow spots (form *glicia*) or is unspotted (form *lativitta*). Female always has pale spots on black forewing border, and yellow (form *glicia*) or white (form *pallida*) ground colour. In form *chrysodona* ground colour is tinted with orange. Spring brood smaller and darker than summer brood. **Nilgiri Clouded Yellow** C. *nilagiriensis* (4.5–5cm; hills of Kerala and Tamil Nadu) is smaller and not subject to much variation – upperside ground colour of males yellow and of females white. **DISTRIBUTION** Austria to Siberia (Sakhalin); North Africa to Himalaya, as far east as central Nepal. **HABITATS AND HABITS** Ascends to 1,200–4,000m in Himalaya. Flight rapid, keeping low over meadows, wheat fields and grassy swathes. Never enters forests, and neither sex has been recorded at wet sand. Both sexes fond of flowers.

Female pallida

Female pallida

Male glicia

Male glicia

Psyche ■ *Leptosia nina* 3.5–5cm

DESCRIPTION Sexes similar. Rounded wings, upperside pattern and small size are singular. **DISTRIBUTION** Throughout India and Sri Lanka to Australia. **HABITATS AND HABITS** Common in dry, open forests at low elevation; ascends Himalaya to 1,500m. Genus is widespread in tropical Africa. Flight weak and low, among bushes in open woodland. Neither sex seems to visit wet sand. Visits low-growing flowers. When it descends with open wings, it is reminiscent of a snowflake and was called thus earlier.

Upperside

Underside

Great Blackvein ■ *Aporia agathon* 8–9cm

DESCRIPTION Sexes similar. Female often has yellowish underside hindwing. Large size, broad dark veins and yellow spot at base of under hindwings distinctive. Western form *phryxe* (Pakistan to western Nepal in inner ranges) paler than eastern forms *caphusa* (central Uttarakhand) and *agathon* (eastern Uttarakhand to North-east India). **DISTRIBUTION** Jammu and Kashmir eastwards to North-east India, Taiwan and Thailand. **HABITATS AND HABITS** Usually found above 1,200m in forests of oak. Flight slow, fluttering and generally about the canopy, but descends to visit thistles and other low-growing flowers and to lay eggs. Single annual brood in April–May, when it can be very common in suitable habitats. Swarms in some years, when clouds of the butterflies congregate on blossoming horse chestnut trees. Mimicked by **Tigerbrown** *Orinoma damaris* in western Himalaya. Blackveins are believed to be distasteful butterflies, but there are records of birds eating them. Males come occasionally to wet mud. Both sexes fond of flowers.

Male upperside caphusa

Underside phryxe

Male underside agathon

Dusky Blackvein ■ *Aporia nabellica* 5–6.5cm

DESCRIPTION Both sexes have prominent dark band across middle of wings, plain on forewing and highly excavated on hindwing. Underside hindwing pale yellow; male pale yellow on upperside, with dark suffusion in forewing cell in subspecies *hesba* (Pakistan), and less so in subspecies *nabellica* (India). Female *hesba* lacks dark suffusion on upperside, while female *nabellica* is densely suffused with dark scales; ground colour in both subspecies varies from yellow to white. **Bhutan Blackvein** *Aporia harrietae* (7–8cm; Bhutan and western Arunachal Pradesh to China) larger, with colour of underside hindwing varying from white to yellow. Upperside suffused with dark scales, except for a few restricted pale marks; underside hindwing has prominent black line between veins. **DISTRIBUTION** Inhabits dry inner ranges from Chitral to Uttarakhand. **HABITATS AND HABITS** Weak flight. Single annual brood in May. Neither sex seems to have been recorded from wet sand. Both sexes visit flowers.

Male nabellica

Bhutan

Female hesba

Female nabellica

Male nabellica

Himalayan Blackvein ■ *Aporia soracta* 5–7cm

DESCRIPTION Sexes similar. Smaller size and narrow black veins differentiate this species from Great Blackvein (see opposite). White underside hindwing and upperside distinguish it from Dusky Blackvein (see above). Similar **Baluchi Blackvein** *A. leucodice* (4–5cm; Baluchistan to Jammu and Kashmir) is smaller, with discal black band on forewing continuous, not broken in middle as in *soracta* (red arrow). **DISTRIBUTION** Jammu and Kashmir to western Nepal. **HABITATS AND HABITS** Occurs in forests of Himalayan Oak above 1,800m. Single annual brood in late May and early June, when the butterfly can be quite common in suitable localities. Swarms about blossoms of horse chestnut and thistles in some years. Great numbers of males fly slowly around low bushes, searching for freshly emerged, unmated females. Males may visit wet mud. Both sexes fond of flowers.

Male underside

Male underside

Male upperside

Large Cabbage White ■ *Pieris brassicae* 6.5–7.5cm

DESCRIPTION Large; forewing apex sharply angled (red circle). Male's upper forewing has no black spots; female has two black spots and black mark along dorsum. Small females can be confused with Indian Cabbage White (see below) in northern India. **DISTRIBUTION** Britain to Japan, North Africa to southern China, throughout Himalaya, winter visitor to Gangetic plain as far south as Delhi. **HABITATS AND HABITS** One of the few butterflies of economic importance, because larvae feed on cabbages. Flight strong, jerky and generally a few metres above the ground. Swarms have been seen migrating in Kashmir. One of the most common butterflies in temperate Himalayan forests in spring. Rarely visits wet sand. Fond of flowers.

Male upperside

Female upperside

Underside male

Underside female

Indian Cabbage White ■ *Pieris canidia* 4.5–6cm

DESCRIPTION Female has two black spots on upper forewing; in male lower spot is almost obsolete. Forewing apex not as sharply angular (red circle) as in Large Cabbage White (see above). **Small Cabbage White** *P. rapae* (4.5–5.5cm; North America, Europe and Asia, to northern Pakistan and Ladakh) smaller, with reduced black apex to forewing. **DISTRIBUTION** Turkistan to Pakistan, east to Singapore and Japan. In India, on southern slopes of Himalaya, generally above 1,000m to more than 3,000m, venturing onto Gangetic plain as far south as Delhi and upper reaches of Nilgiri, Palni and Anamalai hills in southern India. **HABITATS AND HABITS** Flight jerky and erratic, with butterflies settling frequently to bask. Both sexes fond of flowers.

Male upperside

Female upperside

Small Cabbage

Underside

Underside

Western Black-veined White ■ *Pieris melete* 5–6cm

DESCRIPTION Black veins on underside distinguish this group of species. Male upperside has reduced dark suffusion compared with female's. Female yellower than male on both wing surfaces. Similar **Bhutan White** *P. erutae*; Arunachal Pradesh and Bhutan) approximately the same size. **Himalayan White** *P. ajaka* (4–5.5cm) inhabits inner ranges of western Himalaya from Uttarakhand to Pakistan. **DISTRIBUTION** Plains of Assam, Arunachal Pradesh to Myanmar, Korea and Japan. **HABITATS AND HABITS** Occurs at low elevation. Erratic flight, with butterflies generally keeping low and settling frequently on flowers or to bask. Males visit wet sand.

Female underside

LEFT *Male*; RIGHT *Female*

Male underside

Bath White ■ *Pontia daplidice* 4.5–5cm

Male upperside

DESCRIPTION Male has plain white upperside hindwing; female has prominent dusky markings on outer half. On underside, white spot in middle of hindwing small and round, not elongate as in similar **Lesser Bath White** *P. chloridice* (4.5–5cm; high elevation in trans-Himalaya region from Pakistan to Nepal). **DISTRIBUTION** Europe to Japan, along Himalaya from northern Pakistan to Assam, recently extended to Manipur. **HABITATS AND HABITS** Occurs from low elevation to more than 3,500m. Prefers open areas, where its food plant, the North American *Lepidium virginicum*, has colonized roadside swards and cultivation (the plant colonized eastern Himalaya during the last half century). Flight rapid, with butterflies keeping low over sunny meadows, and settling frequently on flowers. Rarely visits wet sand.

Underside

Female upperside

White Orange Tip ■ *Ixias marianne* 5–5.5cm

DESCRIPTION Orange patch on forewing bordered with thick black line. Female has four black spots in orange area on upper forewing. Underside pattern variable, depending upon season, from plain yellow to darker yellow with heavy brown markings. Several similar orange tips, like Plain and Little Orange Tips (see pp. 48 and 49) are immediately distinguished by smaller size and different undersides. **DISTRIBUTION** Endemic to Indian subcontinent, from Pakistan eastwards to West Bengal and southwards to Sri Lanka. **HABITATS AND HABITS** Stragglers ascend hills to nearly 2,000m. Common in open, semi-arid country, and never seen in areas of heavy rainfall or evergreen forests. Flight rapid, near the ground. Males occasionally visit wet sand. Both sexes fond of flowers.

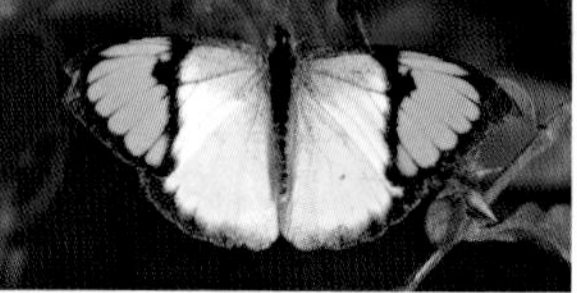

Male upperside

Female upperside

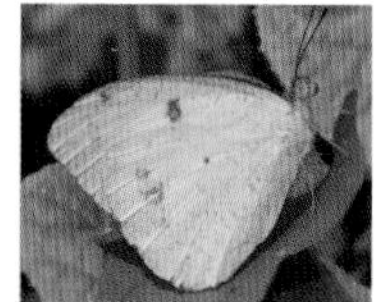

Underside dry season

Underside wet season

Yellow Orange Tip ■ *Ixias pyrene* 5–7cm

DESCRIPTION Male is singular, yellow with bright orange forewing apex. Upperside hindwing dark border variable, and can be very broad in some individuals. Female has broader dark band across forewing, with reduced orange area. Some females lack orange area, having this uniform with rest of wing, either pale yellow or white. Underside seasonally variable, from pale, nearly unmarked yellow in dry season, to darker yellow heavily marked with brown in wet season. Populations in North-east India more heavily marked than their counterparts on peninsula and Sri Lanka. **DISTRIBUTION** Throughout plains of Pakistan, India and Sri Lanka to the Philippines and Indonesia. **HABITATS AND HABITS** Ascends hills to nearly 2,000m. Flight hurried, near the ground. Generally found in open woodland and scrub forest, but also ventures into tropical evergreen forest. Both sexes settle frequently to bask or to visit flowers, while males are fond of wet sand, especially in North-east India, where they form a significant part of large mud-puddling congregations.

Male upperside

Female upperside

Female upperside

Male underside

Large Salmon Arab ■ *Colotis fausta* 4–5cm

DESCRIPTION Sexes similar in North Indian subspecies *fausta*. South Indian female subspecies *fulvia* white. In South India, dark area on upperside forewing margin has three orange or white spots. Can be distinguished from Small Salmon Arab (see below) by larger size and black border on upperside, which is much reduced compared with Small Salmon Arab's. **DISTRIBUTION** North Africa to Asia Minor, through Pakistan as far east as Madhya Pradesh, southwards to drier parts of peninsular India and Sri Lanka. **HABITATS AND HABITS** Flight rapid, and generally low in and among the scrub that comprises its habitat. Settles frequently to bask. Neither sex visits wet sand. Both sexes visit flowers.

Female fulvia

Male fausta

Male fulvia

Small Salmon Arab ■ *Colotis amata* 3.5–5cm

DESCRIPTION Sexes similar, but on upperside entire hindwing of male is bordered with black; in female, dark border begins at apex. On underside, females have a series of discal brown spots across both wings, which males lack. In female form *pallida*, ground colour is white. Distinguished from similar Large Salmon Arab (see above) by smaller size and greater extent of dark markings on upperside of wings. **DISTRIBUTION** Africa through Arabia eastwards to West Bengal, southwards to peninsular India and Sri Lanka. **HABITATS AND HABITS** Inhabits scrubland on both faces of Western Ghats and peninsular India. Active during hottest part of summer days. Neither sex visits wet sand. Visits flowers, but not very fond of them.

Male upperside

Female underside

Female upperside

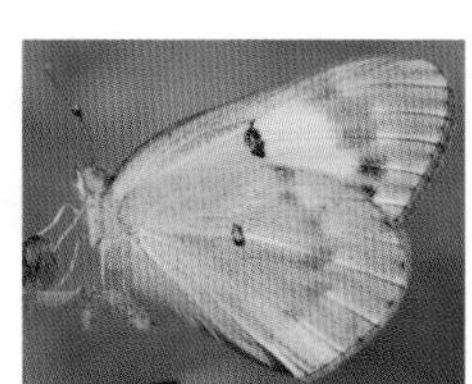

Male underside

Crimson Tip ■ *Colotis danae* 4–5 cm

DESCRIPTION Male has unique crimson tip to forewing. Female duller, with crimson area spotted with black. There are no similar species. **DISTRIBUTION** Africa to southern Pakistan, peninsular India and Sri Lanka. **HABITATS AND HABITS** Inhabits semi-arid regions with low rainfall and scrub forest. Flight rapid, near the ground. Active on the hottest days, when most other butterflies seek shade. Both sexes fond of flowers.

Female upperside

Underside

Male upperside

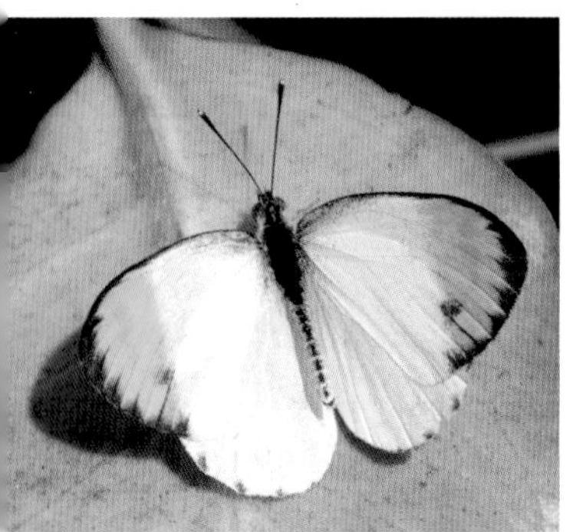

Male upperside

Plain Orange Tip ■ *Colotis aurora* 4–4.5cm

DESCRIPTION Male has orange upper forewing tip, with no dark line separating white and orange parts of wing. Forewing apex black in female, with 3–4 orange or white spots (sometimes absent). Inner edge of black area irregular in this species, while upper half straight in Little Orange Tip (see opposite) female (blue arrow in photo). **DISTRIBUTION** Africa to peninsular India as far east as Madhya Pradesh and southwards to Sri Lanka. **HABITATS AND HABITS** Sometimes common in dry, open areas near its food plants. Flight swift, near the ground. Several individuals gather about flowering bushes and do not leave the area for hours. Active during hottest part of the day.

Female Little Orange Tip

Female upperside

Underside

White Arab ■ *Colotis phisadia* 4–5cm

Upperside

DESCRIPTION Sexes similar. White ground colour and large pale spot in middle of dark forewing border immediately distinguish this species. **Blue-Spotted Arab** C. *protractus* (4–4.5cm; deserts of Sind, Punjab, Gujarat and Rajasthan) has plain underside, pink upperside ground colour, and distinctive blue spots and blue suffusion on upperside. **DISTRIBUTION** Africa to drier parts of northern India as far east as Uttar Pradesh, south to Gujarat and Maharashtra. **HABITATS AND HABITS** Inhabits hot, arid zones. Flight rapid, low among bushes and scrub; active on hot, sunny days. Both sexes stop frequently to visit low-growing flowers.

Blue-Spotted

Upperside Blue-Spotted

Underside

Little Orange Tip ■ *Colotis etrida* 2.5–4.5cm

Male upperside

DESCRIPTION Small. In male red and white areas of forewing separated by straight black band, which in female is curved in lower half (blue arrow). In Plain Orange Tip (see opposite) female, inner edge of dark area is irregular along its length. Female has dark spots on upperside white area, and large red spots below forewing apex, small or absent. **DISTRIBUTION** Throughout plains of Pakistan, India and Sri Lanka; stragglers to 1,800m in Himalaya. **HABITATS AND HABITS** Flight hurried and low. Common in degraded and semi-arid areas. Both sexes fond of flowers.

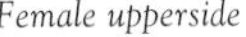
Female upperside

Male underside

Female upperside

Common Albatross ■ *Appias albina* 6–7.5cm

Male upperside

DESCRIPTION Male plain white with narrow black border to forewing apex; distinguished from Striped Albatross (see below) male by upper discocellular vein in forewing cell forming an acute angle, not a right angle (red arrow in figure); female has five white spots on dark area at forewing apex. Ground colour of female may be white or yellow. **DISTRIBUTION** From Sri Lanka to Maharashtra; Nepal eastwards to North-east India, Myanmar and northern Australia. **HABITATS AND HABITS** Found in forested areas. Flight swift and powerful, generally a few metres above the ground. Often flies along forest streams and paths, settling on flowers and circling trees. Males attracted in large numbers to wet sand.

Male underside

Female upperside

Female upperside

Male

Female libythea

Striped Albatross ■ *Appias libythea* 5–6.5cm

DESCRIPTION Male upperside white; subspecies *libythea* differs from Common Albatross (see above) male by vein in forewing cell forming a right angle, *not* an acute angle (red arrow). Female has upperside forewing dark margin with or without white spots. Eastern subspecies *olferna* striped on underside. Wet-season form heavily marked. **DISTRIBUTION** Sri Lanka; peninsular India; Himalaya from Nepal eastwards to north-east India. A migrant, with stragglers recorded from Uttarakahnd. **HABITATS AND HABITS** Flight swift, generally rather high up, but both sexes fond of flowers, and males descend in numbers to wet mud.

Upperside and underside male olferna

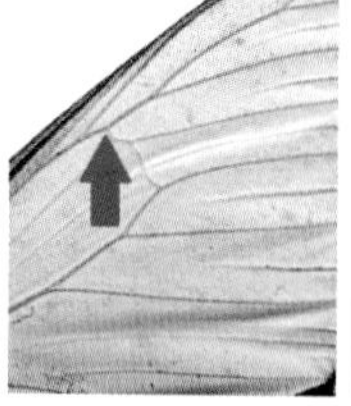
Male libythea

Female wet season olferna

Plain Puffin ■ *Appias indra* 6–7cm

DESCRIPTION Both sexes have no spot in forewing cell and only two white spots on black forewing border. Female distinguished by wider dark borders on upperside, especially on hindwing. Similar **Spot Puffin** *A. lalage* (5.5–8cm; Kerala, Tamil Nadu and from Uttarakhand to Myanmar) has prominent black spot in forewing cell on both surfaces. Female extensively suffused with black on upperside. **DISTRIBUTION** Areas of heavy rainfall in Sri Lanka, western Ghats south of Goa to Kerala, Nepal to Thailand, the Philippines and Borneo. **HABITATS AND HABITS** Low-elevation butterfly, restricted to regions of heavy rainfall. Powerful and jerky flight, generally a few metres above the ground. Female less often seen than male, and usually met in vicinity of evergreen forest. Males gather in large numbers on wet sand. Both sexes visit flowers.

Male upperside

Male underside

Male Spot Puffin

Male underside

Female upperside

Chocolate Albatross ■ *Appias lyncida* 5.5–7cm

DESCRIPTION Yellow under hindwing with chocolate border is singular; female has extensive dark suffusion on upperside. Wet-season form usually more heavily marked than dry-season form, and width of chocolate band on underside varies with season. Some females lack yellow suffusion on underside hindwing. **DISTRIBUTION** Sri Lanka, Maharashtra to Kerala, Nepal to the Philippines and Sulawesi. **HABITATS AND HABITS** Common in forested areas at low elevation. Males usually found in open areas bordering forests, but females are more often seen within forests. Flight rapid and jerky. Males sometimes swarm at wet mud. Both sexes fond of flowers.

Female upperside

Female

Female underside

Male underside

Male upperside

Spotted Sawtooth ■ *Prioneris thestylis* 7–9cm

Male upperside

DESCRIPTION Male's upperside white; female's upperside similar to Hill Jezabel's (see p. 56), but can be distinguished by margin of forewing being straight and 3–4 white spots at end of forewing cell on both surfaces of wing. Dry-season form has reduced dark markings on underside hindwing. **DISTRIBUTION** Along Himalaya from Uttarakhand to North-east India, on to Taiwan and Peninsular Malaysia, generally at low elevation. **HABITATS AND HABITS** A forest insect, usually found bowling along streams and forest paths. Males congregate at wet sand, but females are met within the forest, usually mimicking the slow flight of their models, the Hill Jezabel and its allies. Both sexes fond of flowers.

Female upperside

Male dry season

Spotted sawtooth Female

Pioneer ■ *Belenois aurota* 4–5.5cm

DESCRIPTION Male lightly marked, and female heavily marked with black on upperside. Immediately distinguished by 'J'-shaped black mark above forewing cell (red circle). Dry-season form has white underside; wet-season form under hindwing and tip of forewing chrome yellow in mainland subspecies *aurota* and Sri Lanka subspecies *taprobanis*. **DISTRIBUTION** India and Sri Lanka, extending from West Bengal westwards to Africa, Madagascar, Tajikistan and Afghanistan. **HABITATS AND HABITS** Throughout dry parts in its range. Swift flight. Often migrates northwards in spring in northern India. Both sexes fond of flowers.

Male taprobanis

Female aurota

Male underside aurota

Male underside taprobanis

Lesser Gull ■ *Cepora nadina* 5.5–6.5cm

DESCRIPTION Underside seasonally variable, as shown. Male's upperside white, with broad dark margin on forewing; female's upperside dark with narrow white patches in middle of wing, which are larger in dry-season form. **DISTRIBUTION** Sri Lanka to Gujarat; Nepal to Taiwan and Malaysia. **HABITATS AND HABITS** A forest insect, never found in open country. Found hurrying along forest paths, in glades or along streams. Males visit wet sand. Both sexes fond of flowers.

Male dry season

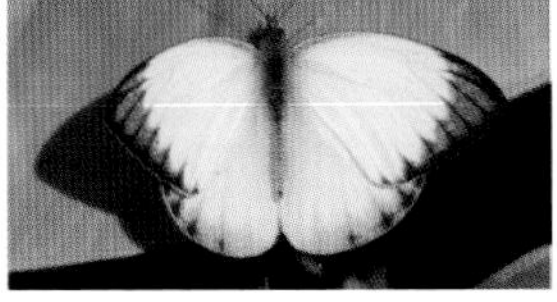

Male upperside

Male wet season

Common Gull ■ *Cepora nerissa* 4–6.5cm

DESCRIPTION Female has heavily marked black veins on upperside; cell on upperside forewing never entirely black. Dry-season form's underside creamy-yellow; wet-season form's bright yellow. Subspecies *evagete* (Rajasthan to Sri Lanka and West Bengal) has white-spotted black forewing apical area; female sometimes has pale yellow upperside spots below forewing apex. In subspecies *phryne* (Himachal Pradesh to North-east India) white spots on black forewing apex of male are reduced or absent, and female never has yellow spots on upperside. Subspecies *lichenosa* (Andaman Islands) has greenish underside hindwing, instead of yellow. **DISTRIBUTION** Throughout Indian mainland from Himachal Pradesh and Rajasthan to Andaman Islands, Taiwan and Sumatra. Appears to be a migrant, but there is no record of where it comes from or goes to in India. **HABITATS AND HABITS** Inhabits open areas or deciduous forests; never found in dense evergreen forests. Swarms in April on North Indian plains. Flight strong and purposeful. Males frequent wet sand. Both sexes fond of flowers, especially lantana.

Male upperside

Male dry season

Male wet season

Female upperside

Red Breast Jezabel ■ *Delias acalis* 8–10cm

Underside

DESCRIPTION Sexes similar. Upperside has red hindwing base. On underside yellow marks on hindwing are elongate and differ in shape from Red Base Jezabel's (see below). **Jezabel Palmfly** *Elymnias vasudeva* (8–9cm; Sikkim eastwards to Myanmar) mimics this species and is found in similar localities, but its shape immediately distinguishes it. **DISTRIBUTION** Himachal Pradesh to North-east India, China and Vietnam. In some years range extends to Uttarakhand and Himachal Pradesh, where it occurs for a few years before disappearing again. **HABITATS AND HABITS** Ascends Himalaya to 1,500m. Inhabits forest in areas of heavy rainfall, where it flies about at treetop level. Distasteful to predators. Males attracted to wet sand. Both sexes fond of flowers.

Male upperside

Jezabel Palmfly

Red Base Jezabel ■ *Delias pasithoe* 7–8.5cm

Male upperside

DESCRIPTION Yellow marks on underside hindwing shorter than in Red Breast Jezabel (see above). No red mark on upperside of wings. Females duller grey above than males. **Red Spot Sawtooth** *Prioneris clemanthe* (8–9cm; Sikkim to Hainan) mimics this species and has similar habits. **DISTRIBUTION** Nepal to North-east India, east to the Philippines and Indonesia. **HABITATS AND HABITS** Common at low elevation in forests and rural landscapes. On the wing within forest, and occurs along paths and streams. Distasteful to predators. Males visit damp sand. Both sexes fond of flowers.

Female upperside

Male upperside

Red Spot Sawtooth

Red Spot Jezabel ■ *Delias descombesi* 8–8.5cm

DESCRIPTION Red spot at base of underside hindwing is elongate. On underside hindwing, male has bright yellow area, which is paler, almost grey, in female. Male and female uppersides as illustrated. **Jezabel Palmfly** *Elymnias vasudeva* (8–9cm; Sikkim to China and Thailand) mimics this species, which can be distinguished at once by shape and somewhat different pattern on wings. **DISTRIBUTION** Nepal to North-east India, Thailand and Indonesia. **HABITATS AND HABITS** Inhabits dense forest at low elevation. Often seen flying along forest paths or streams. Distasteful to predators. Males visit damp sand. Both sexes fond of flowers.

Male upperside

Male underside

Jezabel Palmfly

Female underside

Yellow Jezabel ■ *Delias agostina* 6.5–7cm

DESCRIPTION Underside similar in both sexes, but females may have broader dark veins on forewing. No red mark on underside, which is singular. On upperside, male white with dark suffusion along forewing borders. Female's upper forewing suffused with dark scales. Upperside hindwing similar to underside hindwing, but paler yellow. Mimicked by females of **Jezabel Palmfly** *Elymnias vasudeva* (8–9cm; Sikkim to Thailand), which has different wing shape. **DISTRIBUTION** Nepal to North-east India, Thailand and Vietnam. **HABITATS AND HABITS** Found in dense forests in areas of heavy rainfall, but also ventures onto adjoining open areas and not rarely to rural areas. Flies along forest paths, streams and ridges with a rather strong flight, generally a few metres above the ground. Believed to be distasteful to predators.

Male upperside

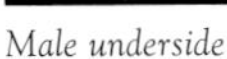

Male underside

Female Jezabel Palmfly

Female upperside

Pale Jezabel ■ *Delias sanaca* 7–8.5cm

DESCRIPTION Sexes similar. Upperside identical to Hill Jezabel but yellow spot at base of hindwing is elongate; ditto on underside. Subspecies *sanaca* (Himachal Pradesh to west Uttarakhand) much paler than eastern subspecies *oreas* (eastern Uttarakhand to Sikkim), and *bhutya* (Bhutan east and southwards). Nearly identical **Dark Jezabel** *D. berinda* (7–8.5cm; Nepal eastwards) can be separated from Pale by examining male genitalia. **DISTRIBUTION** Himachal Pradesh east along Himalaya to China and Vietnam. **HABITATS AND HABITS** Generally a hill insect, inhabiting dense forest at 1,600–2,600m in west Himalaya, lower in east. Flight high around trees, but males readily descend to wet sand. Swarms in some years, when it is so numerous that it is difficult to avoid stepping on it on the streets of hill towns. Distasteful to predators. Both sexes fond of flowers.

Underside sanaca

Underside bhutya

Hill Jezabel ■ *Delias belladonna* 7–8.5cm

DESCRIPTION Sexes similar. Yellow spot at base of forewing is round, not elongate as in Pale Jezabel (see above). West Himalayan subspecies *horsfieldii* (Jammu and Kashmir to Nepal) has wider white-and-yellow markings than east Himalayan *ithiela* (East Nepal to West Bengal at low elevation), or North-east Indian *lugens* (Arunachal Pradesh southwards), which is darkest Indian form. Both sexes very ably mimicked by females of Spotted Sawtooth (see p. 52). **DISTRIBUTION** Along Himalaya from Jammu and Kashmir eastwards to North-east India, Myanmar, China and south to Sumatra. **HABITATS AND HABITS** Ascends Himalaya from low elevation to more than 2,600m. Usually found singly, but swarms in some years. Flight slow, usually high around treetops. Distasteful to predators. Males visit wet sand. Both sexes fond of flowers.

Upperside

Male horsfieldii

Male ithiela

Male lugens

Common Jezabel ■ *Delias eucharis* 6.6–8.3cm

Male upperside

DESCRIPTION Underside pattern distinctive. Red marks along bottom edge of under hindwing heart shaped on black background. Female has darker black veins on upperside. Very similar to **Painted Sawtooth** *Prioneris sita* (8–9cm; endemic to Western Ghats south of Maharashtra and Sri Lanka); red marks on under hindwing rectangular, not heart shaped; sexes similar and upperside similar to Common male's. **DISTRIBUTION** Throughout India and Sri Lanka, eastwards to northern Myanmar. **HABITATS AND HABITS** Ascends to 1,500m. Flies high around trees in search of its larval host plants. Distasteful to predators. Rarely, if ever, visits wet sand. Both sexes fond of flowers and sometimes swarm at lantana bushes.

Painted Sawtooth

Female underside

Painted Jezabel ■ *Delias hyparete* 7–8cm

DESCRIPTION Inner edge of red hindwing-band not bordered with black. Male's upperside white, female's dark. Figured female is of subspecies *mearete* from Singapore. **DISTRIBUTION** Along Eastern Ghats from Tamil Nadu to West Bengal, and along Himalaya from Uttarakhand to North-east India, eastwards to the Philippines and Indonesia at low elevation. **HABITATS AND HABITS** A forest insect that is not as widespread as Common Jezabel '(see above). Flight slow, usually high among trees. Distasteful to predators. Males visit wet sand. Both sexes fond of flowers.

Male upperside

Female upperside

Underside

Pale Wanderer ■ *Pareronia avatar* 6–9cm

Male upperside

DESCRIPTION Male pale blue with unspotted black margin to forewing; female has wide grey margins to both wings. On female's forewing, dark margin has series of white spots that are in line. **DISTRIBUTION** Central Nepal to North-east India and Myanmar. **HABITATS AND HABITS** A forest insect, found in areas of heavy rainfall. Flight erratic, along forest paths and edges. Females mimic Glassy Tiger (see p. 122) in flight. Both sexes fond of flowers.

Female upperside

Male underside

Dark Wanderer ■ *Pareronia ceylanica* 6.5–8cm

Male upperside

DESCRIPTION Male has broad dark borders to wings without pale spots. Female darker than Common Wanderer, (see opposite) with sharper apices to forewing, and submarginal white spots on forewing are reduced compared with female Common's. **DISTRIBUTION** Western Ghats south of Goa to Kerala and Sri Lanka; Andaman Islands. **HABITATS AND HABITS** A forest insect, inhabiting low-elevation, humid forests along seaward face of Ghats. Males visit wet sand. Both sexes fond of flowers.

Female upperside

Male underside

Common Wanderer ■ *Pareronia valeria* 6.5–8cm

DESCRIPTION Both sexes have four pale spots on black margin below forewing apex. Male's ground colour blue; female's white with thicker black marking on veins. White spots on black margin below forewing apex are round. In female of form *philomela* bases of upperside hindwing suffused with yellow. **DISTRIBUTION** Throughout India to the Philippines, except arid parts of Indo-Gangetic plain. **HABITATS AND HABITS** Prefers open forests at low elevation, but also ventures into semi-urban gardens. Male's flight hurried, usually within forest. Female's flight slow, mimicking Glassy Tiger's (see p. 122). Both sexes visit flowers occasionally.

Male upperside

Male underside

Female philomela

Female upperside

Great Orangetip ■ *Hebomoia glaucippe* 8–10cm

DESCRIPTION Large size and pointed forewing apex are singular. Female has dark spots across orange area on upperside forewing, and row of large dark spots along outer edge of hindwing. **DISTRIBUTION** Sri Lanka; forested areas in peninsular India south of Gujarat; Nepal eastwards along Himalaya to Japan and the Philippines. **HABITATS AND HABITS** Prefers areas with moderate to heavy rainfall at low elevation. Flight powerful, often about the canopy of trees. Males sometimes territorial. Underside remarkably resembles a dry leaf when insect is at rest. Both sexes visit flowers and wet sand.

Male upperside

Underside

Underside

Female upperside

Common Punch ■ *Dodona durga* 3–4cm

Male upperside

DESCRIPTION Sexes similar. Pattern on both surfaces of hindwing is singular. **Lesser Punch** *D. dipoea* (3.5–4.5cm; Pakistan to North-east India and western China) distinguished by relatively plain hindwing. **DISTRIBUTION** Pakistan to Nepal, Tibet and China. **HABITATS AND HABITS** Occurs in numbers in humid ravines in dense forest at 1,000–2,500m; males occur singly on sunny ridges. Rapid and skipping flight, with butterflies settling freqently. Males are territorial. Both sexes fond of wet sand and flowers.

Male underside

Male Lesser

Male Lesser

Male upperside

Tailed Punch ■ *Dodona eugenes* 3.5–4.5cm

DESCRIPTION Silvery bands on underside hindwing are distinctive. On upperside, lack of orange distinguishes this from **Orange Punch** *D. egeon* (4.5–5cm; Uttarakhand to North-east India and Malaysia), which is very similar on underside. In both species, females have rounder forewings and larger pale spots. **DISTRIBUTION** Northern Pakistan to North-east India, Taiwan and Malaysia. **HABITATS AND HABITS** Occurs in oak forest at 1,600–2,600m. Flight rapid and skipping. Males are territorial. Both sexes fond of flowers and wet sand.

Male underside

Upperside Orange

Underside Orange

Mixed Punch ■ *Dodona ouida* 4–5.5cm

DESCRIPTION Sexes as illustrated. Underside of male is singular; female can be distinguished from Dark Judy (see overleaf) by shape of hindwing. In **Striped Punch** *D. adonira* (4–5cm; Nepal to North-east India and Java) sexes are similar; underside markings are singular; upperside can be distinguished from Mixed Punch male's by narrower orange bands on forewing. **DISTRIBUTION** Uttarakhand to North-east India, Myanmar and China. **HABITATS AND HABITS** Flight rapid and jerky, and not sustained for long. Males are territorial and take up positions on low trees. Females skulk within forest and are easily confused with treebrowns and judies, which they seem to mimic. Both sexes visit wet sand and flowers.

Female upperside

Underside Striped

Upperside Striped

Male underside

Male upperside

Plum Judy ■ *Abisara echerius* 4–5cm

DESCRIPTION Sexes similar, but females paler than males, with more prominent white band across forewing. There is confusion about the distribution of this species and **Twospot Plum Judy** *A. bifasciata* in India; illustrations of Plum Judy are from Sri Lanka, where Twospot Plum is known not to occur; remaining illustrations are from India. The distinctive characteristic is supposed to be pale band across forewing, which is angled in middle in Twospot Plum and straight in Plum. However, this distinction may not be valid for Indian populations. **DISTRIBUTION** The species occur from Sri Lanka to Gujarat and West Bengal; Himachal Pradesh to North-east India, and further in east Asia. **HABITATS AND HABITS** Occurs in dense forest at low elevation, with stragglers ascending to 1,500m. Males are territorial, patrolling an area a few metres square from a perch on a bush or low tree. Females stay within forest. Both sexes attracted to flowers.

Male

Male

Female Maharashtra

Dark Judy ■ *Abisara fylla* 5–6cm

Male upperside

DESCRIPTION Male has yellow band and female has white band across forewing. Plain underside hindwing with small eyespots distinguishes it from Treebrowns, while rounded wings distinguish it from females of Mixed Punch (see p. 61) and **Tailed Judy** A. *neophron* (4.5–6cm; Nepal to Myanmar and Tibet), which is distinguished by its long hindwing tail. **DISTRIBUTION** Uttarakhand to North-east India and western China. **HABITATS AND HABITS** An inhabitant of dense, humid forests, from low elevation to 2,200m. Flight rapid and jerky, generally settling often. Both sexes fond of flowers and occasionally visit over-ripe fruits.

Female upperside

Underside Tailed

Male underside

Punchinello ■ *Zemeros flegyas* 3.5–4cm

Upperside

DESCRIPTION Sexes similar. Angled hindwing separates Punchinello at a glance from **Harlequin** *Taxila haquinus* (4.5–5cm; Bangladesh to Manipur and Borneo). Some individuals are very dark and others much paler; in Harlequin, females have white bar across upperside forewing. **DISTRIBUTION** Uttarakhand to North-east India, Tibet to Sulawesi. **HABITATS AND HABITS** An inhabitant of dense forest at low elevation in regions of heavy rainfall. Flight rapid and jerky, with butterflies settling frequently. Males are territorial, but are congenial when large numbers visit water together. Both sexes fond of flowers and visit wet sand.

Underside

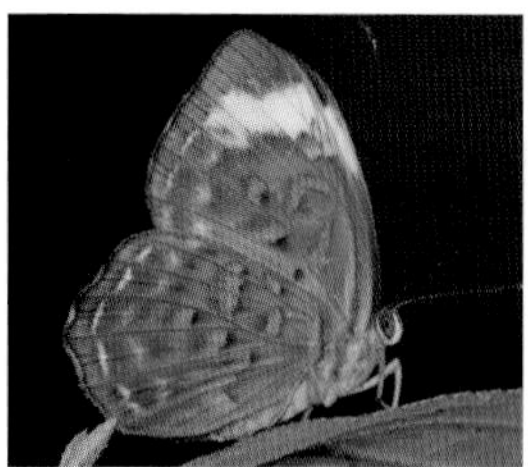
Female Harlequin

Female Harlequin

Indian Sunbeam ■ *Curetis thetis* 4–4.8cm

Male upperside

DESCRIPTION Hindwing rounded and underside not covered with minute black spots. On underside hindwing, dark discal band is more or less continuous across wing, not broken in middle as in **Bright Sunbeam** *C. bulis* and Angled Sunbeam (see below). Males of **Shiva's Sunbeam** *C. siva* (4.6–5cm; Sri Lanka to Karnataka) have tapered hindwing, with underside peppered with minute black spots, but dark band across hindwing is continuous as in Indian. Females difficult to separate from Indian, since the only difference is in black-speckled underside. **DISTRIBUTION** Sri Lanka to Gujarat and eastwards to Odisha. **HABITATS AND HABITS** An inhabitant of low-elevation scrub forests and rural landscapes. Flight rapid but not long sustained. Males are territorial, generally occupying perch on a tree. Both sexes visit water, but do not seem to visit flowers.

Underside Shiva's

Underside

Female upperside

Angled Sunbeam ■ *Curetis acuta* 4–4.5cm

DESCRIPTION Orange/white area extends above base of space 5 on upperside forewing, leaving black 'tooth' in forewing cell (green arrow). On underside, upper portion of discal line is in line with bar at end of cell (both circled in green). In nearly identical **Bright Sunbeam** *C. bulis* (3.5–5cm; Himachal Pradesh Odisha, to North-east India and Japan) orange/white area on forewing does not extend above base of vein 5 on forewing. On underside, discal line is not in line with bar at end of cell. Wing shape variable in both species, less so in Angled. **DISTRIBUTION** Kerala to Gujarat and Odisha; Himachal Pradesh to North-east India, Japan and Thailand. **HABITATS AND HABITS** An inhabitant of scrubby, open areas, where males spend their time patrolling a beat, chasing passing butterflies. Both sexes visit water. Neither species seems to be attracted to flowers, but both occasionally visit over-ripe fruits.

Male upperside

Male Bright

Underside Bright

Male underside

Female upperside

Common Gem ■ *Poritia hewitsoni* 3.1–3.8cm

Male upperside

DESCRIPTION Some females identical to males, others have yellow patch on upper forewing. Black markings on upperside variable. **DISTRIBUTION** Uttarakhand to Malaysia and Vietnam. **HABITATS AND HABITS** Occurs at low elevation, with stragglers ascending Himalaya to 1,500m. Flight rapid and powerful, though females sometimes affect weak flight of the hedge blues (*Celastrina* spp.). They are usually in the forest canopy but descend to water. They do not have any other known attractants. Males are territorial.

Female upperside

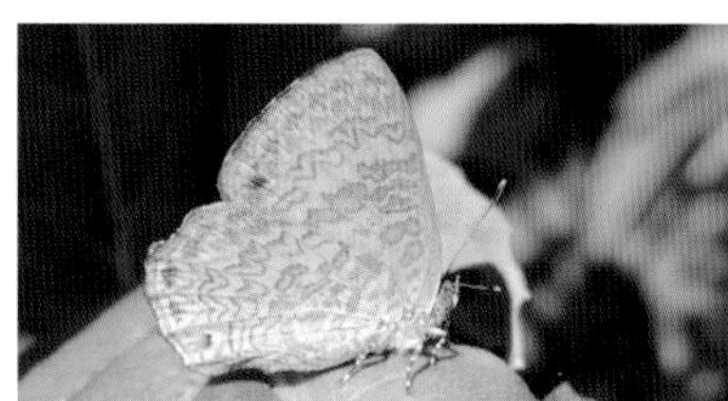

Underside

Common Apefly ■ *Spalgis epius* 2–3cm

DESCRIPTION Sexes similar. Shape and network of fine brown lines on underside are singular. White patch on upperside forewing may be absent. **DISTRIBUTION** Sri Lanka; peninsular India as far north as Gujarat and West Bengal; Uttarakhand to North-east India, the Phillippines and Sulawesi. **HABITATS AND HABITS** Ascends Himalaya to 1,500m and appears wherever its larval food, scale insects, occurs. Flight rapid and males are often territorial. Neither sex has been recorded at flowers. Upperside resembles some forms of Malayan (see p. 85).

Female underside

Male upperside

Female upperside

Common Copper ■ *Lycaena phlaeas* 2.6–3.4cm

DESCRIPTION Sexes similar, but females generally have brighter orange on upperside forewing than males. Row of blue spots may be present on hindwing. Similar **White Bordered Copper** (*L. panava*, 3.7–4cm; Pakistan to central Nepal) distinguished by white line bordering red band on underside hindwing; upperside forewing of males never suffused with dark scales. **DISTRIBUTION** Throughout Europe, North and East Africa, temperate Asia, North America and along Himalaya. **HABITATS AND HABITS** Occurs on sunny, open hillsides from 1,000m to well over 4,000m. Flight brisk and not sustained. Both sexes visit flowers and generally found in immediate vicinity of their larval host plant, *Rumex*.

Underside

Male upperside

White Bordered

White Bordered

Sorrel Sapphire ■ *Heliophorus sena* 2.8–3.3cm

DESCRIPTION Sexes similar, but females usually have prominent orange spots on upperside forewing border. Underside is singular. **DISTRIBUTION** Pakistan to Arunachal Pradesh. Previously known from western Himalaya, but recently reported from western Arunachal Pradesh. **HABITATS AND HABITS** Occurs at 400–2,700m. Flight rapid but not long sustained. Butterfly settles frequently to bask and visits flowers. Usually quite common in immediate vicinity of its food plant, *Rumex hastatus*.

Underside

Male upperside

Female upperside

Golden Sapphire ■ *Heliophorus brahma*

DESCRIPTION Male is golden. On upperside hindwing female has faint grey line between tornus and tail (red arrow). **DISTRIBUTION** Uttarakhand to North-east India, China and Thailand. **HABITATS AND HABITS** Occurs at 1,200–2,500m. Flight rapid, over low-growing bushes where it settles frequently to bask. Males are territorial and both sexes are fond of flowers.

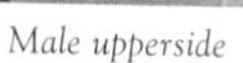
Male upperside

Female upperside

Underside

Eastern Blue Sapphire ■ *Heliophorus oda* 3.2–3.6cm

DESCRIPTION Sexes as illustrated. Lacks tails on hindwing. Distinguished from **Western Blue Sapphire** *H. bakeri* (3–3.4cm, Chitral to western Nepal) by presence of dark discal line across underside forewing (red arrow), which is reduced to series of dashes or lacking in Western. **DISTRIBUTION** Himachal Pradesh to central Nepal. **HABITATS AND HABITS** A very local butterfly found in glades and on banks of streams in oak forest above 1,600m. Flight weak and near the ground, and males pugnaciously attack other males. Both sexes fond of flowers.

Male upperside

Male upperside Western

Female upperside

Female underside

Western

Common Silverline ■ *Spindasis vulcanus* 2.6–3.4cm

Upperside

DESCRIPTION Underside markings consist of red bands with black edges. On underside forewing, mark in cell (vertical blue arrow) consists of streak and spot. On underside hindwing, sub-basal band never crosses vein 1b (horizontal blue arrow). Upperside has prominent orange area on forewing and there is no blue iridescence. Similar **Plumbeous Silverline** *S. schistacea* (2.8–3.7cm; Sri Lanka to Kerala and Gujarat) distinguished by male having iridescent blue scales on upperside hindwing, and female having greyish scales on upperside hindwing and at base of forewing. **DISTRIBUTION** Sri Lanka, throughout India to Thailand, except arid regions. **HABITATS AND HABITS** Ascends hills to 2,500m. Flight rapid but not sustained. Males are territorial and both sexes are fond of flowers.

Underside Plumbeous

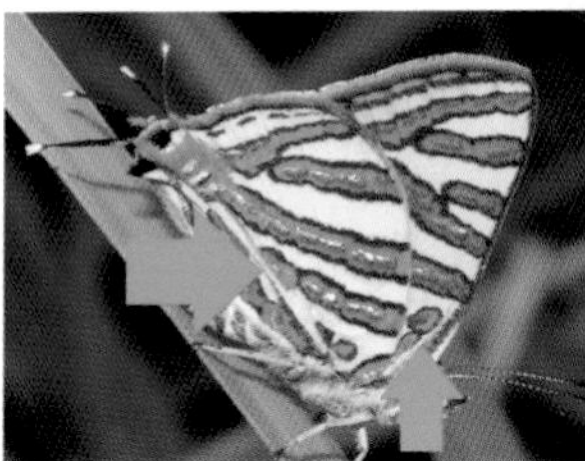
Underside

Long Banded Silverline ■ *Spindasis lohita* 3.6–4.2cm

DESCRIPTION Sexes similar. Underside colour and pattern very variable, but sub-basal band usually extends to orange area near tail (vertical black arrow). On underside forewing, sub-apical bands (horizontal black arrow) usually form 'V'. Upperside lacks orange on forewing. **DISTRIBUTION** Sri Lanka, Maharashtra to Kerala; Uttarakhand to North-east India, Taiwan and Indonesia. **HABITATS AND HABITS** Ascends Himalaya to 1,600m. Flight rapid, with butterflies settling frequently. Males are territorial and attracted to wet sand. Both sexes fond of flowers.

Underside

Underside

Male upperside

Common Shot Silverline ■ *Spindasis ictis* 2.7–3.5cm

Spring form

DESCRIPTION Upper forewing has well-marked, triangular orange patch, and in male iridescent blue area does not extend above vein 2. Underside ground colour variable from dark red to pale yellow. On hindwing, sub-basal band is broken into three spots (vertical red arrow and red circle). In **Scarce Shot Sliverline** *S. elima* (2.8–4.2cm; Pakistan, Sri Lanka and India) iridescent area extends above vein 2 on upperside forewing, and female has oval or circular orange patch, never triangular. **DISTRIBUTION** Sri Lanka, throughout India. **HABITATS AND HABITS** Generally found in open country. Flight rapid, with butterflies settling often. Males are territorial and active on hottest days. They visit wet sand. Both sexes fond of flowers.

Male upperside

Underside

Underside

Silver Hairstreak ■ *Inomataozephyrus syla* 4.2– 4.5cm

DESCRIPTION Male and female as illustrated. Underside ground colour silvery with sub-marginal line not ending in large dark spot near tornus. In **White Spotted Hairstreak** *Shizuyaozephyrus ziha* (3.5–3.8cm; Himachal Pradesh to Uttarakhand) sexes are similar, and on underside forewing there are two prominent dark spots at tornus. **DISTRIBUTION** Afghanistan to Pakistan eastwards to western Nepal. **HABITATS AND HABITS** An inhabitant of dense oak forest at 1,500–2,700m. Flight weak and fluttering, with butterflies settling frequently at water. Neither sex attracted to flowers. Swarms in some years during May–June.

Male underside

White Spotted

Male upperside

Female upperside

White Spotted

Wonderful Hairstreak ■ *Thermozephyrus ataxus* 4–4.6cm

DESCRIPTION Unusually, undersides of sexes differ. Large size and red spots on upperside forewing of female distinguish the species from others. **DISTRIBUTION** Pakistan to North-east India, Myanmar to Taiwan and Japan. **HABITATS AND HABITS** Occurs in oak forests at 1,200–2500m. Flight swift, generally about treetops, but both sexes descend to water (I recorded a male near water in June at 11.30 p.m.). Neither sex known to visit flowers. Though generally not frequently seen, there are population outbreaks in some years.

Male upperside

Male underside

Female underside

Female upperside

Large Oakblue ■ *Arhopala amantes* 4.5–5.7cm

DESCRIPTION Underside hindwing has iridescent greenish scales near hindwing tail. Male's upperside shining blue with narow black border. Female's upperside forewing has broad dark border notched at end of cell. Similar **Centaur Oakblue** A. *centaurus* (5.3–6.2cm; Sri Lanka north to Maharashtra, Uttarakhand to North-east India to the Philippines and Indonesia) distinguished on underside by vertical silvery lines in forewing cell, and lack of any iridescent greenish scales near hindwing tail. **DISTRIBUTION** Sri Lanka to Gujarat and Odisha, Uttarakhand to North-east India, Malaysia and Timor. **HABITATS AND HABITS** An inhabitant of dense forest at low elevation, with stragglers ascending to 1,200m in Himalaya. Flight rapid and erratic, generally about the canopy, but both sexes descend to water. Members of this genus often gather in great numbers about a particular tree, where they settle on leaves and twigs during the day.

Male above; female below

Male upperside Centaur

Male upperside Centaur

Dark Himalayan Oakblue

■ *Arhopala rama* 3.4–4cm

DESCRIPTION Male and female as illustrated. Dark blue upperside and dark brown underside distinguish this species from **Pale Himalayan Oakblue** A. *dodonea* (3.8–4.4cm; north Pakistan to central Nepal; sexes similar), which is much paler. **DISTRIBUTION** North Pakistan to North-east India, Myanmar and Thailand. **HABITATS AND HABITS** Inhabits dense oak forest at 1,200–2,500m. Flight rapid and jerky, with butterflies generally staying around the canopy. Both sexes come to water. Never found in large numbers. Active throughout the year.

Male upperside

Female upperside

Upperside Pale Himalayan

Underside Pale Himalayan

Underside

Indian Oakblue ■ *Arhopala atrax* 3.4–4cm

DESCRIPTION Male as illustrated, female with broader black forewing border. Pale brown underside with relatively long hindwing tail is distinctive. **DISTRIBUTION** Peninsular India; Jammu and Kashmir to North-east India and Malaysia. **HABITATS AND HABITS** An inhabitant of dense forest at low elevation, with rare stragglers at 1,200m. Swarms during summer in *Shorea robusta* forest, when hundreds may be seen gathered along stream beds. Flight swift but not sustained. Neither sex has been recorded at flowers.

Male upperside

Underside

Tailless Bushblue ■ *Arhopala ganesa* 3.2–3.7cm

DESCRIPTION Sexes similar. Rounded hindwing and smaller size distinguish it from Pale Himalayan Oakblue (see opposite). **Dusky Bushblue** *A. paraganesa* (3–3.4cm; Uttarakhand to Borneo) distinguished by tails on hindwing. **DISTRIBUTION** Kashmir to North-east India, north Myanmar to Japan. **HABITATS AND HABITS** Inhabits oak forest at 1,000–2,500m. Flight rapid and jerky. Both sexes visit water. Males occasionally territorial but readily gather at mud-puddling assemblages with dozens of other males.

Upperside

Underside

Dusky

Common Acacia Blue ■ *Surendra quercetorum* 3–4cm

DESCRIPTION Male's upperside shining blue; female's brown. Shape of wings is singular. Male has one tail, female two tails on hindwing. Males of South Indian population lack any blue on upperside hindwing. Genus has two species in India, this one and **Burmese Acacia Blue** *S. vivarna* (Sri Lanka to Gujarat; Andaman and Nicobar Islands), but there is no clear superficial difference between them and it is unclear whether they occur together in some parts of the range. **DISTRIBUTION** Forested areas along Himalaya from Uttarakhand to North-east India. **HABITATS AND HABITS** Rarely visits flowers or water; fond of basking on prominent perches.

Male upperside

Female Burmese

Male Burmese

Female underside

Purple Leaf Blue ■ *Amblypodia anita* 4.5–5.2cm

DESCRIPTION Male and female as illustrated. Large size and wing shape are singular. Pattern and shade of underside vary from white to dark brown. **DISTRIBUTION** Sri Lanka to Gujarat, eastwards to North-east India and Indonesia. **HABITATS AND HABITS** An insect of dense forest and always keeps to shade. Settles frequently on leaves. Males visit water.

Male upperside

Female upperside

Underside

Silverstreak Blue ■ *Iraota timoleon* 4–4.8cm

Male upperside

DESCRIPTION Male's upperside ground colour shining blue, female's dark blue. Underside is singular in Sri Lanka and peninsular India. South-east of Assam **Scarce Silverstreak Blue** *I. rochana* (4–4.8cm, Assam; Meghalaya; Manipur to Sulawesi) occurs, which has larger silver streak across underside hindwing. **DISTRIBUTION** Sri Lanka to Gujarat and Uttar Pradesh; Himachal Pradesh to North-east India and Malaysia. **HABITATS AND HABITS** Stragglers ascend Himalaya to 2,400m. Flight very swift but not sustained. Males are territorial and visit water. Both sexes occasionally visit flowers.

Female upperside

Underside

Yamfly ■ *Loxura atymnus* 3.6–4cm

DESCRIPTION Sexes similar. Some individual variation in width of dark border to wings on upperside. **Branded Yamfly** *Yasoda tripunctata* (3.2–4cm; at low elevation from North-east India to Vietnam and Thailand) has much the same habits. **DISTRIBUTION** Dry forests in Sri Lanka and India eastwards to Indonesia and the Philippines. **HABITATS AND HABITS** Ascends to 1,500m. Flight not very strong, and often near the ground. Settles frequently in undergrowth, on twigs or prominent leaves. Some males take up a beat a metre above the ground, moving from perch to perch within an area of a few square metres. Has not been recorded at water.

Underside Branded

Upperside Branded

Male dry season

Male wet season

Underside mating

Common Onyx ■ *Horaga onyx* 2.7–3.3cm

DESCRIPTION Sexes similar. Female with duller blue than male. Similar **Brown Onyx** *H. viola* (2.2–2.8cm; Sri Lanka to Goa and Himachal Pradesh to North-east India) lacks all blue on upperside and has white band across underside hindwing reduced to a line. **DISTRIBUTION** Sri Lanka to Maharashtra; Himachal Pradesh to North-east India and Indonesia. **HABITATS AND HABITS** A forest insect that ascends to 2,000m. Although males are not territorial, the species is found singly. Generally stays high among treetops and does not remain in one place for long.

Upperside

Underside

Brown

Monkey Puzzle ■ *Rathinda amor* 2.6–2.8cm

DESCRIPTION Sexes similar. Hindwing has three tails. White line along edge of underside forewing is distinctive. Dry-season form paler than wet-season form. **DISTRIBUTION** Sri Lanka through peninsular India to North-east India. **HABITATS AND HABITS** Found in clearings in forests at low elevation. Males are territorial, generally occupying a patch of sunlight within a dense forest, patrolling bushes and shrubs. Does not seem to visit wet sand. Both sexes visit flowers.

Male underside

Female underside

Upperside

Common Imperial ■ *Cheritra freja* 3.8–4.2cm

DESCRIPTION Sexes similar. Male has purple tinge to brown upperside. In **Blue Imperial** *Ticherra acte* (3.4–3.8cm; Uttarakhand to North-east India and Borneo), male has two hindwing tails and female has three, one long and two short. Both sexes have two white spots above two black tornal spots on upperside hindwing. On underside, dry-season form is orange and wet-season form is pale brown. **DISTRIBUTION** Sri Lanka to Maharastra; Uttarakhand to North-east India and Indonesia. **HABITATS AND HABITS** Found in dense forest at low elevation in areas of heavy rainfall. Males are territorial during the morning hours, and usually take up a beat around a bush. Even a small patch of sunlight within a forest is enough for a male to patrol. Neither sex appears to be fond of wet sand or flowers.

Underside

Female

Male

Upperside Blue

Underside Blue

Common Tit ■ *Hypolycaena erylus* 3.2–3.8cm

DESCRIPTION Male's upperside blue with distinctive large black patch in middle of forewing. Female as illustrated. On underside, ochreous line across middle of wings is distinctive. **DISTRIBUTION** Central Nepal to North-east India, the Philippines and New Guinea. **HABITATS AND HABITS** An inhabitant of dense forest at low elevation in areas of heavy rainfall. Flight rapid but not sustained. Males are territorial, usually taking up position on a bush. They occasionally visit wet sand.

Underside

Male upperside

Female upperside

Chocolate Royal ■ *Remelana jangala* 3.2–4.2cm

DESCRIPTION Dark chocolate underside of male is singular, while female has paler ground colour on both surfaces of wing. **DISTRIBUTION** Central Nepal to North-east India, the Philippines and Sulawesi. **HABITATS AND HABITS** Found at low-elevation and dense forest, in regions with heavy rainfall. Males visit wet sand, but females rarely seen away from forest. Males take up a beat on bushes. Both sexes occasionally visit flowers.

Male upperside

Underside

Underside Orchid

Blue Tit ■ *Chliaria kina* 2.6–3.1cm

DESCRIPTION Sexes similar, but female has broader dark borders on forewing than male, and blue area is paler. Similar **Orchid Tit** *C. othona* (2.4–2.7cm; Maharashtra to Kerala; Uttarakhand to North-east India and Borneo) has a different underside. Male's upperside as illustrated, female's brown. **DISTRIBUTION** Uttarakhand to North-east India; Taiwan to Thailand. **HABITATS AND HABITS** An inhabitant of dense humid forest, ascending hills to 2,400m. Flight ragged and insect settles frequently. Males attracted to water and females rarely seen.

Upperside Orchid

Male upperside

Underside

Cornelian ■ *Deudorix epijarbas* 3.4–4.4cm

Underside diara

DESCRIPTION Male's upperside ground colour red, female's brown. Patterns of lines of underside are distinctive, with bar at end of forewing cell (red arrow) very near discal band on both wings unlike in Common Guava Blue (see opposite). Ground colour variable, as in aberration *diara*. **DISTRIBUTION** Sri Lanka to peninsular India; Pakistan to North-east India, south to Australia. **HABITATS AND HABITS** Ascends to 2,600m in Himalaya. Butterfly of wooded country. Flight very swift but long sustained. Males visit wet sand. Both sexes fond of flowers. Larvae feed inside fruits and may cause damage in pomegranate orchards.

Male upperside

Female upperside

Underside

Fluffy Tit ■ *Zeltus amasa* 2.8–3.2cm

DESCRIPTION Two long, fluffy tails on hindwing, the longer one originating from tornus, are distinctive. Male's upperside as illustrated, while female's is brown. **DISTRIBUTION** Goa to Kerala; Central Nepal to Indonesia. **HABITATS AND HABITS** An inhabitant of dense forest in areas of heavy rainfall. Males territorial, taking up position on bushes. Flight rapid and darting. Females remain within the forest and are rarely seen. Males come readily to wet sand, but neither sex visits flowers.

Underside

Male upperside

Common Guava Blue
■ *Virachola isocrates* 3.4–5cm

DESCRIPTION Male iridescent blue above; female brown with orange patch on forewing. Underside pattern distinctive, with bar at end of cell well separated from discal band on both wings (see Cornelian, opposite for comparison). **DISTRIBUTION** Throughout India and Sri Lanka to northern Myanmar. **HABITATS AND HABITS** Occurs at low elevation, ascending Himalaya to 2,000m. Flight very swift. Males are territorial. Both sexes fond of flowers. Larvae feed on fruits of guavas and tamarind.

Male upperside

Underside

Female upperside

Plane ■ *Bindahara phocides* 3.6–4.2cm

DESCRIPTION Male's underside ground colour brown, female's white; male's upperside black with iridescent blue spots on hindwing, female's brown. Area above upperside hindwing tail pale brown in male and white in female. **DISTRIBUTION** Sri Lanka to Goa and Sikkim to Australia. **HABITATS AND HABITS** An inhabitant of humid evergreen forest at low elevation. Rarely found outside forest. Males are territorial on treetops, but descend readily to flowers. Flight rapid but not long sustained.

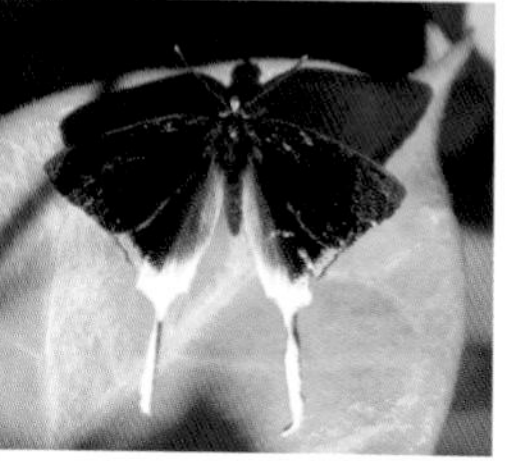

Male upperside

Male underside

Indian Red Flash ■ *Rapala iarbus* 3.3–4.1cm

DESCRIPTION Male's upperside bright red and veins may be black. Female's duller red to brown with diffused black borders. On underside bands are narrow. Similar **Scarlet Flash** *R. dieneces* (3–3.6cm; West Bengal to the Philippines and Borneo) has broader black border to forewing; male's upperside red and female's brown. Yellowish underside ground colour. **DISTRIBUTION** Sri Lanka, throughout India to Borneo. **HABITATS AND HABITS** Low-elevation butterfly with stragglers ascendng to 2,400m in Himalaya. Flight rapid, settling frequently on prominent leaves and twigs to bask. Males are territorial. Both sexes fond of flowers.

Male upperside

Underside Scarlet

Female upperside

Underside

Scarlet

Slate Flash ■ *Rapala maena* 3–3.3cm

Underside Indigo

DESCRIPTION Sexes similar, but male has iridescent blue on upperside and female is duller. On underside dark bands, whose width varies individually, are bordered with white only on outer side, unlike in **Indigo Flash** *R. varuna* (2.8–3.2cm; Himachal Pradesh to Sri Lanka and Australia), where dark bands are often much wider and bordered on both sides with white. On upperside, not iridescent blue like Slate. **DISTRIBUTION** Throughout India and Sri Lanka to Borneo. **HABITATS AND HABITS** Occurs from low elevation to 2,200m. Flight rapid and not long sustained. Males are territorial. Both sexes fond of wet sand and flowers.

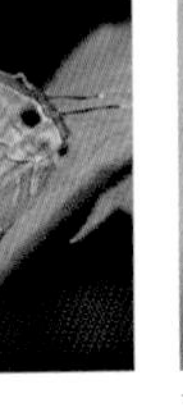
Underside

Male upperside

Upperside Indigo

Common Lineblue ■ *Prosotas nora* 1.8–2.5cm

DESCRIPTION Sexes similar. Hindwing with thread-like tail. On forewing band nearest to base extends below cell (black arrow). Underside very variable. Male's upperside forewing has narrow dark border; female's forewing has broad dark border. **Tailless Lineblue** *P. dubiosa* (2.2–2.6cm; throughout Sri Lanka, India to Australia) lacks hindwing tail; on upperside the sexes are similar to Common. Cilia of both wings plain brown. **White-tipped Lineblue** *P. noreia* (2.2–2.8cm; Sri Lanka to Uttarakhand and North-east India onwards to Indonesia) has white tip of forewing cilia, and on underside forewing band nearest to base does not extend beyond cell. **Dingy Lineblue** *Petrelaea dana* (2.4–2.8cm; Sri Lanka to Uttarakhand and eastwards to New Guinea) is tailless, has brown cilia and two black spots at bottom of hindwing are almost equal in size. **DISTRIBUTION** Sri Lanka to northern India, eastwards to Australia. **HABITATS AND HABITS** Flight brisk and difficult to follow. Males congregate in large numbers at wet sand. Both sexes visit flowers.

Dingy

Tailless

White-tipped

Underside

Angled Pierrot ■ *Caleta decidia* 2.6–3.2cm

DESCRIPTION Sexes similar. Angled black band at base of forewing distinguishes this and **Elbowed Pierrot** *C. elna* (3–3.4cm; Odisha to central Nepal and eastwards to Sulawesi) from other members of the genus. In Elbowed basal band on hindwing is usually broad and upperside is black. **Banded Blue Pierrot** *Discolampa ethion* (2.6–3cm; Sri Lanka to Gujarat; Uttarakhand to Sulawesi) has double band at base of underside forewing; male's upperside blue and female's brown. **DISTRIBUTION** Sri Lanka to Gujarat and Odisha; western Nepal to Vietnam. **HABITATS AND HABITS** Occurs at low elevation in forested regions with moderate to heavy rainfall. Males congregate at wet sand and both sexes are attracted to flowers. Strong on the wing. Males are sometimes territorial around bushes and small trees.

Upperside

Underside

Banded Blue

Elbowed

Dark Cerulean ■ *Jamides bochus* 2.5–3.4cm

DESCRIPTION Can be distinguished from lineblues by lack of any markings on basal half of underside forewing. Upperside of male is a unique iridescent blue. Female's upperside paler, with broad dark border to forewing. **DISTRIBUTION** Throughout Sri Lanka and India to Japan and Sulawesi. **HABITATS AND HABITS** Butterfly of forested hills ascending to 2,600m. Flight brisk and long sustained. Males visit wet sand. Both sexes fond of flowers. Rarely found in large numbers, though several may be seen at the same time around their larval host plants.

Male upperside

Female upperside

Underside

Common Cerulean ■ *Jamides celeno* 2.7–4cm

DESCRIPTION On upperside male has narrower black border on forewings than female. On underside dry-season form has broad brown bands and wet-season form has bands bordered by prominent white lines. Both forms are shown mating in the photograph. On underside forewing third vertical white line from base extends across wing (red arrow). **DISTRIBUTION** Sri Lanka to Gujarat; north Pakistan to North-east India and Sulawesi. **HABITATS AND HABITS** Ascends Himalaya to 1,500m. Inhabits woodland and forest. Flight weak and fluttering. Settles frequently on leaves. Males occasionally visit wet sand. Both sexes visit flowers.

Male and female upperside

LEFT *Wet season*; RIGHT *Dry season*

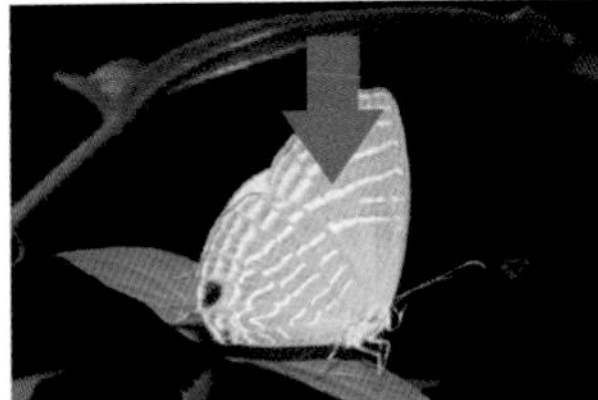

Male upperside

Forgetmenot ■ *Catochrysops strabo* 2.5–3.5cm

DESCRIPTION Tailed. On underside forewing, round spot on costa is well separated from discal band (red arrow). Upperside ground colour of male pale blue. In **Silver Forgetmenot** *C. panormus* (2.5–3.5cm; Sri Lanka to Karnataka; central Nepal to north Australia) upperside of male is silvery-blue, and on underside forewing of both sexes, spot on costa is near discal line (compare with Forgetmenot). Uppersides of females of both species identical. **DISTRIBUTION** Throughout Sri Lanka and India, and Sulawesi. **HABITATS AND HABITS** Migrant species found in all habitats. Flight rapid and long sustained. Males visit wet sand. Both sexes visit flowers.

Male Silver

Male upperside

Female upperside

Underside

Silver

Peablue ■ *Lampides boeticus* 2.4–3.6cm

DESCRIPTION Underside pattern, especially broad white band on outer half of wing, is singular. Male and female as illustrated. **Zebra Blue** *L. plinius* (2.2–3cm; Sri Lanka to north Pakistan and eastwards to Australia) has a somewhat different pattern, especially on underside forewing. On upperside, Peablue has two black spots near hindwing tail that are lacking in male Zebra Blue. Female Zebra Blue very different from female Peablue.

Zebra Blue

Male underside

DISTRIBUTION Europe to Russia and South Africa; Arabia through India to Australia and Hawaii. **HABITATS AND HABITS** A migrant found almost throughout the Old World. Migrates in large numbers in spring in Himalaya. Flight strong and sustained. Males visit wet sand. Both sexes visit flowers.

Male upperside

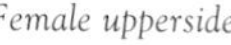

Female upperside

Male upperside Zebra Blue

Common Pierrot ■ *Castalius rosimon* 2.4–3.2cm

DESCRIPTION Sexes similar. Female often more heavily marked than male. Underside is singular. Wet-season form more heavily marked than dry-season form. **Forest Pierrot** *Taraka hamada* (2–3cm; Sikkim to China and Borneo) similar, but tailless with four black marks on underside along forewing costa. Male's upperside is brown, while female's is white with broad dark costa and margin. **DISTRIBUTION** Sri Lanka to northern India and eastwards to Sulawesi. **HABITATS AND HABITS** Found in forest and open woodland, with stragglers ascending to 2,400m in Himalaya. Flight weak and fluttering, generally in undergrowth. Males congregate in large numbers on wet sand. Both sexes fond of flowers.

Male upperside

Underside

Underside Forest

Pale Grass Blue ■ *Pseudozizeeria maha* 2.6–3cm

DESCRIPTION Ground colour of underside brownish-grey. On underside there is a spot in forewing cell (red arrow). Black spots on hindwing often duller than those on forewing. Male's upperside pale blue, female's brown, sometimes with some blue suffusion at base of wing. Underside of **Dark Grass Blue** *Zizeeria karsandra* (1.8–2.4cm; North Africa through Indian subcontinent and Sri Lanka to Australia) identical to Pale's, except that ground colour is pale grey and black marks on hindwing are as prominent as marks on forewing. On upperside, male Dark is dark blue with broad dark border, and female is brown. **DISTRIBUTION** Iran eastwards through Indian subcontinent and Tibet to Korea. **HABITATS AND HABITS** Ascends to 2,400m in Himalaya. A butterfly of degraded areas where its larval hostplant, *Oxalis*, grows. Flight weak and fluttering, close to the ground and not long sustained. Males fond of wet sand. Both sexes fond of flowers.

Dark

Female upperside

Male upperside

Male upperside

Dark

Lesser Grass Blue ■ *Zizina otis* 1.9–2.6cm

DESCRIPTION Both this and **Tiny Grass Blue** *Z. hylax* (1.6–2.4cm; throughout Africa and from Arabia through Indian subcontinent and Sri Lanka to Australia) lack spot in underside forewing cell (compare with Pale, see above). Lesser differs from Tiny in spot on hindwing being under costal spot (red arrow); spot is shifted outwards in Tiny. **DISTRIBUTION** Pakistan to Sri Lanka, Japan and Sulawesi. **HABITATS AND HABITS** Widespread species, found in all types of country. Flight relatively strong and sustained, often several metres above the ground. Females found in vicinity of their larval host plants. Males attracted to wet sand. Both sexes fond of low-growing flowers.

Tiny

Male upperside

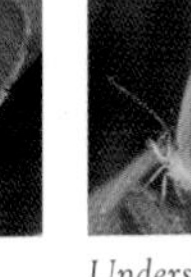

Female upperside

Underside

Tiny

Red Pierrot ■ *Talicada nyseus* 2.8–3.6cm

DESCRIPTION Sexes similar. Upperside and underside are singular. Subspecies *nyseus* occurs in most of India and Sri Lanka, while subspecies *khasiana* is found in North-east India and north Myanmar. Distinguished by smaller orange area on upperside hindwing. **DISTRIBUTION** Throughout India except eastern coast and Bihar. Has extended range to northern India and western Himalaya during this century. **HABITATS AND HABITS** Flight weak and fluttering, and species is generally found in immediate vicinity of its succulent host plants, within whose leaves the butterfly spends its early stages.

Male upperside

Female upperside

Underside

Quaker ■ *Neopithecops zalmora* 2–3cm

DESCRIPTION Sexes similar. Wings rounded. Upperside brown with central white patch that may be absent or may occupy almost whole wing. Underside has sparse dark markings that are seasonally variable. **DISTRIBUTION** Sri Lanka to Gujarat and Odisha; Jammu and Kashmir to North-east India and Timor. **HABITATS AND HABITS** A forest insect that is common in areas of high humidity. Ascends to 1,500m. Within the forest flight is weak and fluttering, with butterflies settling frequently on leaves, low-growing flowers or wet mud.

Male above; female below

Underside

Underside

Malayan ■ *Megisba malaya* 2–3cm

DESCRIPTION On underside forewing costa there are 4–6 dark marks. Upperside white patch of varying extent. Sri Lankan and southern Indian subspecies *thwaitesi* lacks tail on hindwing. Both tailless and tailed forms occur in Odisha, Arunachal Pradesh and Singapore. **Common Hedge Blue** *Acytolepis puspa* (2.7–3.5cm; Afghanistan to Sri Lanka, Japan and New Guinea) similarly marked on underside, but lacks costal spots on forewing. **DISTRIBUTION** Sri Lanka to Maharashtra and West Bengal; Uttarakhand to North-east India and New Guinea. **HABITATS AND HABITS** Found in dense forest at low elevation, with stragglers ascending to 1,500m. Within forest flight is weak and butterfly settles frequently. Males attracted to wet sand. Both species visit flowers.

Common Hedge Blue

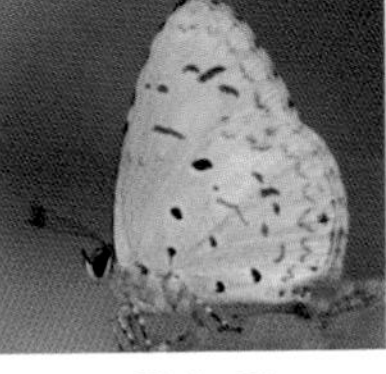

Common Hedge Blue

Male upperside

Underside

Gram Blue ■ *Euchrysops cnejus* 2.5–3.3cm

DESCRIPTION Both sexes have two black spots at bottom of hindwing. On underside, spot in space 1b is distant from next spot in space 1c (green circle and red arrow); distinguished from Plains Cupid (see p. 86) by lack of spot in space 1b. **Small Cupid** *Chilades parrhasius* (2–2.5cm; Sri Lanka to Assam, westwards to Arabia) has underside hindwing spot in space 1b adjacent to spot in space 1c. On upperside, male has one black hindwing spot and female is brown. **DISTRIBUTION** Throughout Pakistan, India and Sri Lanka to Australia. **HABITATS AND HABITS** Inhabitant of open, scrubby areas, where it feeds on legumes. Larvae sometimes cause damage to gram crops. Flight rapid and long sustained. Males come to wet sand. Both sexes fond of flowers.

Male upperside

Male Small Cupid

Female upperside

Underside Small Cupid

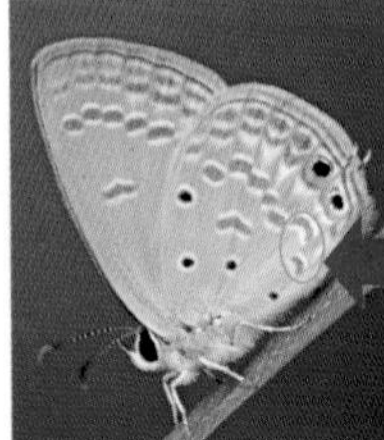

Underside

Plains Cupid ■ *Luthrodes pandava* 2.5–3.5cm

DESCRIPTION Male's upperside pale blue, female's with broad dark borders. On underside hindwing, spot at base of space 1b (red circle) is always present and distinguishes this species from Small Cupid and Gram Blue (see. p. 85). **Lime Blue** *Chilades laius* (2.6–3cm; throughout India, Pakistan and Sri Lanka to the Philippines) lacks tail on hindwing. On upperside, both sexes similar to Plains Cupid, but females have reduced spot at bottom of hindwing; pattern on underside is seasonally variable. **DISTRIBUTION** Throughout India and Sri Lanka. **HABITATS AND HABITS** Ascends to 2,200m in Himalaya. Resident of open, scrubby lands. Weak flyer and settles frequently. Males gather in numbers at wet sand, and both sexes visit water.

Underside

Underside Lime Blue

Female upperside

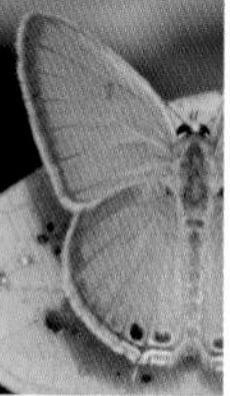

Male upperside

Male Lime Blue

Grass Jewel ■ *Freyeria trochylus* 1.5–2cm

Male upperside

DESCRIPTION Sexes similar. Hindwing has three prominent black spots that may or may not be crowned with orange. Very similar **Lesser Grass Jewel** *F. putli* (1.5–2.2cm; Sri Lanka to Uttarakhand and eastwards to Australia) has four black spots along edge of hindwing. **DISTRIBUTION** Throughout Africa to Europe, Azerbaijan through India to Thailand. **HABITATS AND HABITS** A butterfly of open woodland and scrubland. Flies weakly about low-growing plants, settling frequently on leaves and flowers. Males gather at wet sand.

Male underside

Underside Lesser

Upperside Lesser

Brown Awl ■ *Badamia exclamationis* 5–5.5cm

DESCRIPTION Very narrow forewings and shape of hindwing are distinctive. Female has extra white spot on forewing. **DISTRIBUTION** Throughout Sri Lanka and India, to Japan and New Guinea. **HABITATS AND HABITS** Flight very swift, and the insect does not stay in one place for long. Settles under leaves. Both sexes visit flowers and males visit bird droppings.

Underside

Upperside

Common Banded Awl ■ *Hasora chromus* 4.5–5cm

DESCRIPTION White band on hindwing is narrow and relatively faintly marked. Female has additional pale spot on forewing. **Orange-tailed Awl** *Bibasis sena* (4.5–5cm; Sri Lanka to Maharashtra and Madhya Pradesh; Himachal Pradesh to North-east India and Sulawesi; Andaman and Nicobar Islands) has wider band across hindwing, and bottom of hindwing is orange. **White Banded Awl** *Hasora taminatus* (4.5–5.5cm; Karnataka to Sri Lanka; Himachal Pradesh to North-east India and New Guinea; Andaman and Nicobar Islands) has wide white band across hindwing and lacks orange. Male's upperside is unmarked, but female has some pale spots on forewing. **DISTRIBUTION** Throughout Pakistan, India and Sri Lanka to Australia. **HABITATS AND HABITS** A common species in city parks and pristine forests, ascending Himalaya to more than 2,000m. Flight swift and difficult to follow. Both sexes fond of flowers.

Male upperside

Common Banded

White Banded

Orange-tailed

Orange Awlet ■ *Burara jaina* 6–7cm

DESCRIPTION Underside hindwing has narrow orange stripes between veins; in Himalayan subspecies *jaina*, underside forewing has pale spots, which may be faintly marked in some individuals; these are lacking in southern Indian subspecies *fergussoni*. Males of **Branded Orange Awlet** *B. oedipodea* (4–7cm; Sri Lanka; Andaman Islands; Himachal Pradesh to North-east India and Sulawesi) have large black patch on upperside forewing, which may be reduced or broken into spots in some populations; distinguished from Orange Awlet on underside by pale, diffuse orange patches in middle of both wings and near bottom of hindwing. **DISTRIBUTION** Maharashtra to Kerala; Himachal Pradesh to North-east India and Taiwan. **HABITATS AND HABITS** An inhabitant of dense evergreen forests in regions of heavy rainfall. Ascends hills to around 2,000m. Active at dusk, when males are territorial, patrolling beats a few metres in extent about 3m above the ground. Flight powerful and skipping. Males visit wet sand early in the morning. During daytime, both sexes visit flowers, though they are largely inactive during this period.

Branded Orange

Male jaina

Female fergussoni

Branded Orange

Tricoloured Pied Flat ■ *Coladenia indrani* 3.5–4cm

DESCRIPTION Sexes similar, but on underside female has prominent yellow spots on hindwing, which are relatively smaller in male. Upperside hindwing has prominent dark spots, which are sharply defined in Himalayan subspecies *indrani*, and less so in illustrated southern Indian subspecies *indra*. Subspecies *indrani* is tawny, while both *indra* and Sri Lankan subspecies *tissa* are dark brown with some tawny marks along edges of wing. **DISTRIBUTION** Sri Lanka to Gujarat and West Bengal; Himachal Pradesh to North-east India and Laos. **HABITATS AND HABITS** Occurs at low elevation in dense forests, in regions of heavy rainfall. Flight rapid, and insect settles frequently on undersides of leaves.

Himalayan White Flat ■ *Seseria dohertyi* 4.5–5cm

DESCRIPTION Sexes similar. Abdomen mostly white, with short dark tip. In addition, along inner margin of forewing there is a white streak that is lacking in **Sikkim White Flat** *S. sambara* (4–4.5cm; Sikkim to northern Vietnam). Sikkim also has much narrower white band across abdomen, leaving almost half of lower part dark. **DISTRIBUTION** Uttarakhand to northern Vietnam. **HABITATS AND HABITS** An inhabitant of dense forests with heavy rainfall. Ascends Himalaya to 2,000m. Flight is swift. Males attracted to wet sand. Both sexes very fond of flowers. Active during daytime and in full sun.

Upperside

Upperside Sikkim White

Common Yellow-breast Flat ■ *Gerosis bhagava* 3.5–4.5cm

DESCRIPTION Sexes similar, but abdomen is pale yellow in males and white in females. Similar **Dusky Yellow-breast Flat** *G. phisara* (3.5–4.5cm; central Nepal to Hainan) suffused with dusky scales, and has narrower white band and abdomen striped with narrow white bands, on every segment of abdomen in females, and on middle three segments in males. **DISTRIBUTION** Kerala to Goa and Jharkhand; central Nepal to North-east India and Thailand. **HABITATS AND HABITS** An inhabitant of dense forests at low elevation in regions of heavy rainfall. Flight is swift. Usually seen at wet sand, where it can spend up to half an hour imbibing liquid. Little is known of its habits. Both sexes visit flowers.

Upperside

Upperside Dusky

Large Snow Flat ■ *Tagiades gana* 4.5–5.5cm

DESCRIPTION Both sexes individually variable. Some individuals have white suffusion on hindwing and others lack it. Forewing has three translucent specks near apex, but there are never any other translucent specks or spots on other parts of wings. **Suffused Snow Flat** *T. japetus* (4.5–5cm; Kerala to Maharashtra; central Nepal to Borneo) has one or more translucent spots in middle of forewing, and is also individually variable, in terms of white suffusion on hindwing. **Common Snow Flat** *T. parra* (Central Nepal to Borneo) has some dark spots along bottom of hindwing on white area. **DISTRIBUTION** Kerala to Maharashtra; central Nepal to the Philippines and Borneo. **HABITATS AND HABITS** Found in dense forest at low elevation in regions of heavy rainfall. Flight swift and powerful, generally within forest. Settles frequently on leaves. Both sexes fond of flowers.

Upperside

Upperside Common

Upperside Suffused

Water Snow Flat ■ *Tagiades litigiosa* 3.7–4.4cm

DESCRIPTION Sexes similar. Black spots on hindwing can be large or small, depending upon season. Distinguished from **Spotted Snow Flat** *T. menaka* (4.3–5.5cm; Jammu and Kashmir to Myanmar to China and Thailand) by lack of two black spots in white area of hindwing (red arrows). **Striped Snow Flat** *T. cohaerens* (4.4–4.6cm; Himachal Pradesh to North-east India and Thailand) distinguished by its black-and-white striped abdomen. **DISTRIBUTION** Sri Lanka to Maharashtra and Jharkhand; Andaman Islands; Himachal Pradesh to North-east India and Indonesia. **HABITATS AND HABITS** A forest insect, not uncommon at low elevation in regions of heavy rainfall. Butterfly active all day, with males patrolling small glades or patches of sunlight within forest. Males known to visit bird droppings and wet sand. Both sexes fond of water and flowers.

Upperside dry season

Upperside wet season

Upperside Spotted

Fulvous Pied Flat ■ *Pseudocoladenia dan* 3.5–4cm

DESCRIPTION Sexes similar. Upperside hindwing does not have sharply defined black marks, and underside does not have yellow spots, as in Tricoloured Pied Flat (see p. 88). **Sikkim Pied Flat** *P. fatua* (3.5cm; Sikkim to North-east India and Tibet) is smaller and redder, and forewings are slightly pronounced. **DISTRIBUTION** Kerala to Gujarat and Andhra Pradesh; Himachal Pradesh to North-east India and Sulawesi. **HABITATS AND HABITS** Ascends Himalaya to more than 1,500m. Inhabitant of forested areas, usually seen patrolling small, sunny patches or glades. Flight rapid and skipping, always near the ground. Settles frequently on low-growing shrubs and flowers.

Sikkim upperside

upperside

Spotted Small Flat ■ *Sarangesa purendra* 2.5–3.5cm

DESCRIPTION Sexes similar. Small size distinguishes this and **Common Small Flat** *S. dasahara* (3–3.5cm; Sri Lanka to Rajasthan and Odisha; Himachal Pradesh to Vietnam) from almost all other skipper species. Pale spots on forewing distinguish Spotted from Common, which lacks them. **DISTRIBUTION** Rajasthan to Kerala; Himachal Pradesh to Uttarakhand. **HABITATS AND HABITS** Flight skipping and butterfly keeps within 1m of the ground, usually flying about over its larval host plants. Males take up low-level beat along forest paths and in clearings. Both sexes fond of flowers. They are active throughout the day and favour sunny patches.

Underside

Upperside

Upperside Common

African Marbled Skipper ■ *Gomalia elma* 2.5cm

DESCRIPTION Sexes similar. Small size and unusual pattern are singular. **DISTRIBUTION** Sri Lanka to Telangana, Maharashtra and Himachal Pradesh, westwards through Pakistan to southern Arabia and throughout Africa and Madagascar. **HABITATS AND HABITS** An inhabitant of open grassland in dry regions that has managed to colonize even small clearings in forests. Flight rapid and darting, low over grass and herbs, settling frequently on low-growing flowers. Has been seen to hold forewings over hindwings like a moth, adding to strong resemblance of wing pattern to a moth's.

Upperside

Underside

Indian Skipper ■ *Spialia galba* 2.5cm

DESCRIPTION Small size and unique chequered pattern are distinctive, and species can be distinguished from other butterflies of its size at a glance. **DISTRIBUTION** Throughout Pakistan, India and Sri Lanka to Hainan and Thailand. **HABITATS AND HABITS** No special habitat preferences and can be found almost anywhere, ascending to 2,500m. Small colonies occur in suitable habitat and are active throughout the day, unlike most members of this family. Butterflies keep low over grass and herbs, settling frequently to bask or on low-growing flowers.

Upperside

Underside

Tiger Hopper ■ *Ochus subvittatus* 2.2–2.7cm

DESCRIPTION Sexes similar. Upperside dark brown, with yellow band across upper end of forewing that may be absent. Underside singular. **DISTRIBUTION** Uttarakhand to North-east India, Thailand and China. **HABITATS AND HABITS** An insect of dense forest at low elevation, found in small colonies. Flight not very powerful, and butterfly is active during the hours of daylight. Males are territorial, taking up a beat from a leaf of a bush or low-growing tree in a forest glade, or along a forest path. Returns repeatedly to the same perch. Males visit wet sand.

Mating pair

Male underside

Chestnut Bob ■ *Iambrix salsala* 2.6–3cm

DESCRIPTION Upperside brown with variable number of pale spots. Ground colour of underside hindwing, along with two central spots, is singular and serves to immediately distinguish this species from other skippers. **DISTRIBUTION** Sri Lanka to Gujarat, Uttarakhand to North-east India, China and Indonesia. **HABITATS AND HABITS** A forest insect that is restricted to humid regions. Usually found within forests, where males take up a beat along a forest path or glade. They perch on a leaf of a bush or low-growing tree, and investigate passing butterflies. Both sexes occasionally visit flowers.

Upperside

Female underside

Male underside

Grass Demon ■ *Udaspes folus* 4–4.8cm

Upperside

DESCRIPTION Sexes similar. Upperside and underside patterns unique. **DISTRIBUTION** Sri Lanka to Gujarat and West Bengal; Himachal Pradesh to North-east India, Japan and Indonesia. **HABITATS AND HABITS** An inhabitant of forests in regions of heavy rainfall, but also in more open places where its larval host plants, species of ginger, grow. Occurs on hills to more than 2,000m. Flight rapid and not very high above the ground. Males visit wet sand. Both sexes fond of flowers, and occasionally visit dung.

Upperside

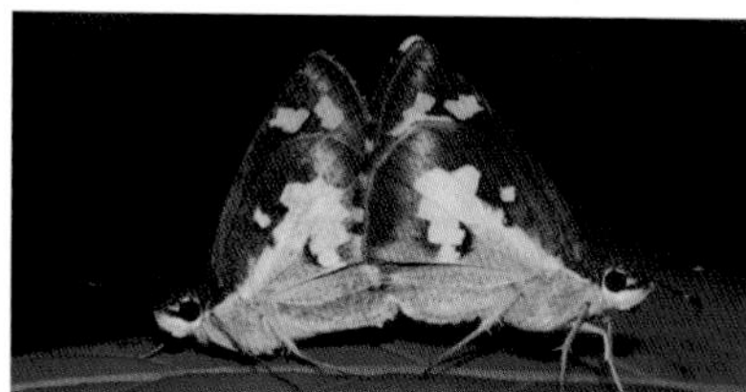
Mating pair

Indian Palm Bob ■ *Suastus gremius* 3.2–4.2cm

DESCRIPTION Sexes similar. Series of dark spots on underside hindwing is singular; these spots may sometimes be rather faintly marked. **DISTRIBUTION** Throughout Sri Lanka and India to Taiwan and Malaysia. **HABITATS AND HABITS** Low-elevation species found wherever palms, its larval host plants, grow. Flight swift and generally high, but both sexes come readily to flowers. Males visit damp sand and bird droppings. They do not seem to be territorial.

Upperside

Underside

Banana Skipper ■ *Erionota torus* 7–7.7cm

DESCRIPTION Sexes similar. Large size immediately distinguishes this from most other skippers. Distinguished from closest relative, **Palm Redeye** *E. thrax* (7–7.7cm; Karnataka to Kerala; Sikkim to North-east India and New Guinea) by slightly straighter outer margin of forewing, which is slightly more rounded in Palm Redeye. From the only other skipper of similar size, **Giant Redeye** *Gangara thyrsis* (7–7.6cm; Sri Lanka to Maharashtra and Andhra Pradesh; Himachal Pradesh to North-east India and Sulawesi) both the former species differ in lacking three small pale spots below forewing apex. **DISTRIBUTION** Sri Lanka; Tamil Nadu, Karnataka, Kerala; Uttarakhand to North-east India, Malaysia and Hawaii. **HABITATS AND HABITS** Has population outbreaks on plantations of banana, its larval host plant. Flight swift, generally about 3m above the ground. On the wing at dusk.

Underside Giant Redeye

Underside

Male upperside

Common Redeye ■ *Matapa aria* 4–4.5cm

DESCRIPTION Sexes similar. Upperside unmarked brown; male has obscure black brand. Cilia on hindwing paler than ground colour, varying from grey to pale yellow. In North-east India there are several similar species, but in Sri Lanka, Indian peninsula and western Himalaya this species is unique. **DISTRIBUTION** Sri Lanka to Gujarat, Gangetic plain, Uttarakhand to North-east India and the Philippines. **HABITATS AND HABITS** Flight swift, at level of bushes. Generally a forest insect, but visits gardens and open areas. Males are territorial and both sexes are fond of flowers. While they have not been recorded at wet sand, they do tend to suck up moisture on leaves of bushes.

Underside

Underside

Male upperside

Yellow Coster ■ *Acraea vesta* 4.5–7cm

DESCRIPTION Sexes rather similar, but most of upperside in male is unmarked, while in female yellow area is marked with heavy dark lines. **DISTRIBUTION** Himachal Pradesh to Myanmar, occupying belt at 1,200–2,600m. **HABITATS AND HABITS** Found both in forests and open country, wherever its larval food plants grow. There are outbreaks of this species on chosen bushes, and it swarms in the vicinity in due course. Flight weak and fluttering, and it is a local butterfly, never found far from its food plant. Voted the ugliest butterfly in India, it contains poisons – African members of the genus are known to contain prussic acid in their body tissue. Both sexes fond of flowers.

Male upperside

Female upperside

Underside

Tawny Coster ■ *Acraea violae* 5–6.5cm

DESCRIPTION Sexes similar. Narrow wings and row of white spots on black hindwing border make this a singular butterfly in India. **DISTRIBUTION** Sri Lanka; throughout India east to Singapore, which it has recently colonized. **HABITATS AND HABITS** Flight weak and fluttering, generally in vicinity of its food plant, but occasionally stragglers ascend to 2,400m in Himalaya in dispersal flights. Both sexes fond of flowers. It is apparently quite poisonous, for it is rarely attacked by predators. African members of the genus contain prussic acid (hydrogen cyanide) in their body tissue. Sharing a feature with snow apollo butterflies (*Parnassius* spp.), mated female has tip of abdomen plugged by mate to prevent mating with other males.

Upperside

Underside

Red Lacewing ■ *Cethosia biblis* 6.5–9cm

DESCRIPTION Both sexes lack pale discal band across forewing. Ground colour of female varies from red, like male's, to greenish-blue (illustrated). **DISTRIBUTION** Nepal along Himalaya to the Philippines and Sulawesi. **HABITATS AND HABITS** Usually occurs at low elevation, but ascends eastern Himalaya to more than 2,000m. Flight slow and leisurely and individuals are often local, circling a bush or clearing for hours. This and other members of the genus are believed to be protected by distasteful chemicals in the body tissue. Both sexes fond of flowers.

Male upperside

Female underside

Male underside

Female upperside

Leopard Lacewing ■ *Cethosia cyane* 8.5–9.5cm

DESCRIPTION Broad white band across forewing and striking pattern on underside immediately distinguish this species from others. **DISTRIBUTION** Uttarakhand eastwards to North-east India, Odisha to Thailand and Vietnam. **HABITATS AND HABITS** Generally found at low elevation, but stragglers have been recorded from up to around 2,000m in west Himalaya. While it is common in North-east India, it appears to be a straggler in west Himalaya and there are very few records from there. Species is protected from predators by poisons in body tissue, and flies leisurely to advertise this.

Underside

Female

Male

Tamil Lacewing ■ *Cethosia nietneri* 8–9.5cm

DESCRIPTION Males of Sri Lankan race *nietneri* lack almost all of red on wings, and white band across forewing is highly reduced. Females overlaid with dark scales and entirely lack broad white band across forewing typical of Indian race *mahratta*. **DISTRIBUTION** Endemic to Sri Lanka and Western Ghats from Maharashtra to Kerala. **HABITATS AND HABITS** Flight leisurely, usually in open areas bordering evergreen forests. Both sexes fond of flowers. Believed to be protected from predators by poisons contained in body tissue.

Male nietneri

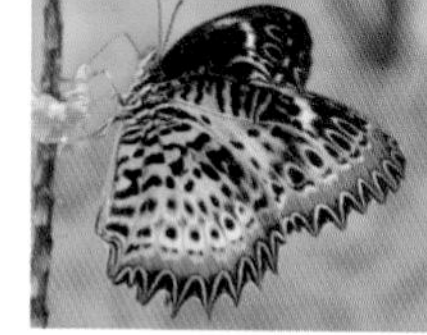
Male nietneri

Male mahratta

Female mahratta

Underside mahratta

Common Beak ■ *Libythea lepita* 4.5–5cm

DESCRIPTION Sexes similar. Upperside has orange streak, divided into streak and round spot in Himalayan subspecies *lepita*, while streak is divided into two, making a total of three marks, in southern Indian and Sri Lankan subspecies *lepitoides*. Extraordinarily long palps give this and the next species the common name of beaks. **European Beak** *L. celtis* (4–5cm; Chitral to Europe) distinguished by wider orange markings. **DISTRIBUTION** Pakistan along Himalaya to Myanmar. **HABITATS AND HABITS** One of the earliest butterflies on the wing in Himalaya, emerging from February onwards. Flight generally high around branches of its larval host tree – usually a flap-and-glide progression. When it settles on leafless twigs of its larval host plant, *Celtis* spp., its underside makes it look remarkably like a dead leaf. Males are territorial, though they take breaks to visit flowers or wet sand.

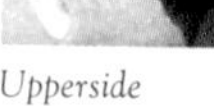
Upperside

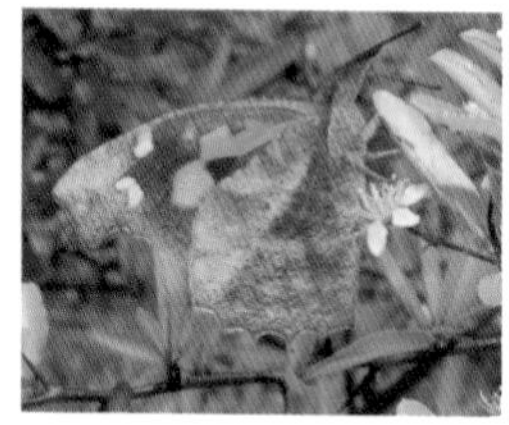
Underside

Club Beak ■ *Libythea myrrha* 4.5–5.5cm

Underside

Male sanguinalis

DESCRIPTION Sexes similar. Long palpi are distinctive in members of this genus. Orange band on upper forewing undivided and club shaped. Southern Indian subspecies *carma* is smaller than Himalayan subspecies *sanguinalis*. Females usually have wider orange bands than males. **DISTRIBUTION** Sri Lanka; Western Ghats south from Gujarat. Himalaya and North-east India to Borneo and Palawan. **HABITATS AND HABITS** A forest insect, ascending Himalaya to more than 2,400m. Uses flap-and-glide technique to fly. Among the earliest butterflies on the wing in spring in Himalaya. Both sexes fond of flowers and wet sand.

Female sanguinalis

Male carma

Black Prince ■ *Rohana parisatis* 4.5–5cm

DESCRIPTION Male all black. South Indian subspecies *atacinus* and Sri Lankan subspecies *camiba* have three white specks below costa of forewing; Himalayan subspecies *parisatis* has only one white speck on forewing. Female somewhat resembles castors (*Ariadne* spp.), but has fewer bands across forewing and on underside, which closely resembles male's. **DISTRIBUTION** Sri Lanka to Maharashtra; Himalaya from Nepal to North-east India, Borneo and Palawan. **HABITATS AND HABITS** Normally at low elevation, but ascends to 1,800m in southern India. Flight swift, with males often patrolling small patch of territory to waylay passing females. Females have slower flight, rather like flap-and-glide technique of castors, but are capable of faster flight if the need arises. Males attracted to wet sand.

Female underside

Male underside

Male parisatis

Female atacinus

Sordid Emperor ■ *Chitoria sordida* 6–7cm

DESCRIPTION Sexes similar. Colour and pattern on wings, especially single pale band across forewing, are singular, except for **Naga Emperor** *C. naga* (6–7cm; Nagaland to Thailand), which has yellowish band. **DISTRIBUTION** Sikkim to Vietnam. **HABITATS AND HABITS** An inhabitant of forests in regions of heavy rainfall up to 2,000m. Both sexes fond of over-ripe fruits and tree sap, and may be found in orchards adjoining forests. Flight powerful, comprising a few wingbeats followed by a glide.

Male upperside

Female underside

Naga underside

Indian Purple Emperor ■ *Mimathyma ambica* 6.5–7.5cm

Underside

DESCRIPTION Male has iridescent blue upperside; female's is brown, with rounder wings. Similar **Chitral Emperor** *M. chitralensis* (7.5–9cm; Kashmir to Chitral) is larger and duller, with brown band on underside hindwing lacking black dot in lower half. **DISTRIBUTION** From Kashmir along Himalaya to China and Sumatra. **HABITATS AND HABITS** Occurs generally at 1,500–2,500m in the west, at low elevation in eastern Himalaya. Flight swift and powerful, with flap-and-glide technique when not disturbed. Males visit wet sand, and both sexes visit over-ripe fruits, rotting animal carcasses and dung.

Male upperside

Female upperside

Golden Emperor *Dilipa morgiana* 7–8cm

DESCRIPTION Male has golden-bronze markings, female the same markings in white. Stout body and short wings set this butterfly apart from all other Indian butterflies. **DISTRIBUTION** Kashmir along Himalaya to northern Vietnam. **HABITATS AND HABITS** Flight powerful, often around favourite treetops, but both sexes readily descend to feed on over-ripe fruits or even rhododendron blossoms in spring, when there are no fruits. In the morning, males can be seen patrolling low over wet grass, settling to suck up moisture.

Male underside

Male upperside

Eastern Courtier *Sephisa chandra* 7.5–9cm

DESCRIPTION Male has white outer band of spots on upperside forewing. Female has three forms: in form *chandra* orange markings replaced with blue-and-black ones (underside illustrated); form *albina* has same forewing markings as male, but they are white, as is hindwing (illustrated); form *chandrana* is like typical form, but outermost row of white spots is large, occupying most of wing below apex. **DISTRIBUTION** Nepal to North-east India, Thailand and Taiwan. **HABITATS AND HABITS** Flight powerful and males are strongly territorial. Females mimic flight of their models, the blue crows (*Euploea* spp.). Both sexes visit water and are attracted to over-ripe fruits. When there is no fruit, they sometimes visit flowers.

Male upperside

Male underside

Female underside chandra

Female upperside albina

Western Courtier ■ *Sephisa dichroa* 6–7.5cm

DESCRIPTION Sexes similar; female has rounder wings. Markings on forewing are orange, which distinguishes this species from similar Eastern Courtier (see p. 101). **DISTRIBUTION** Pakistan (Chitral) to western Nepal; south-east China. **HABITATS AND HABITS** Along Himalaya, common in summer and autumn in forests of Himalayan Oak, its larval host plant. Flight powerful, generally high among treetops. Both sexes are very fond of over-ripe fruits, wet mud and dung, at which dozens of the insects can gather on sunny days.

Male upperside

Female upperside

Male underside

Male underside

Pasha ■ *Herona marathus* 7–9cm

DESCRIPTION Sexes similar. Wing pattern and colour similar to those of orange-and-black sailers (*Neptis* spp.) and sergeants (*Athyma* spp.), but forewing shape is unique. **DISTRIBUTION** Eastern Nepal to Myanmar, Thailand and Vietnam. **HABITATS AND HABITS** Flight moderately strong and generally high, but both sexes descend to over-ripe fruits and tree sap, of which they are very fond. Females flutter about along forest edges, at times looking very like female sergeants.

Male upperside

Female underside

Painted Courtesan ■ *Euripus consimilis* 6–8.5cm

DESCRIPTION Males have distinctive row of red spots along edge of hindwing, which females lack. A very distinctive species. **DISTRIBUTION** Maharashtra to Kerala; Uttarakhand to North-east India and Thailand. **HABITATS AND HABITS** A low-elevation, forest butterfly, occasionally ascending to 1,500m. Flight weak and fluttering, very reminiscent of a zygaenid moth. Both sexes attracted to flowers and over-ripe fruits, and males occasionally visit wet sand.

Female upperside

Male underside

Male upperside

Courtesan ■ *Euripus nyctelius* 6.5–8.5cm

DESCRIPTION Male rather like Painted Courtesan (see above), but lacks all red marks. Distinguishable from Siren (see p. 104) by irregular shape of hindwing; on forewing, outermost row of white dots does not reach apex as it does in Siren. Female Courtesan very variable, with five named forms in India: typical form looks like form *isa* but has narrow dark borders to hindwings; form *nyctelius* has iridescent blue upperside forewing; form *alcathoeoides* lacks blue iridescence of *nyctelius*; form *cinnamoneus* resembles *nyctelius*, but has row of white streaks along hindwing. **DISTRIBUTION** Sikkim to North-east India, the Philippines and Indonesia. **HABITATS AND HABITS** A low-elevation, forest insect, whose several female forms closely mimic blue and brown crows (*Euploea* spp.). Both sexes visit flowers and over-ripe fruits. Male has a rather strong flight, but females share slow flight of their models.

Male upperside

Female nyctelius

Male underside

Female isa

Siren ■ *Hestina persimilis* 6.5–7.5cm

DESCRIPTION Eyes are pale brown. Males more heavily marked with black than females, though pattern is the same. Hindwing more or less rounded, with slight projection in middle of outer margin. **DISTRIBUTION** Himachal Pradesh along Himalaya to North-east India, Odisha to western China. **HABITATS AND HABITS** Flight of males rapid, consisting of series of fast wingbeats followed by a glide. They generally stay about the canopies of trees, sometimes taking up a position to display territorial behaviour. Females mimic leisurely flight of Glassy Tiger (see p. 122), and are sometimes difficult to distinguish at first glance. Both sexes fond of over-ripe fruits, though they sometimes also visit flowers when fruits are lacking.

Female underside

Female upperside

Circe ■ *Hestinalis nama* 9.5–10.5cm

DESCRIPTION Sexes similar, but female has more rufous area on upper hindwing than male. Good mimic of Chestnut Tiger (see p. 123), but can be easily separated by black marks in forewing cell. Larger **Yellow Kaiser** *Penthema lisarda* (12.3–13.5cm; Sikkim to North-east India and Hainan) lacks tawny suffusion on hindwing. **DISTRIBUTION** Uttarakhand to North-east India, east to Hainan and south to Indonesia. **HABITATS AND HABITS** Occurs from low elevation to more than 2,000m, more or less within distribution limits of its model, the Chestnut Tiger. A forest insect, with neither sex venturing far into cultivated areas. Flight of males is swift and powerful, settling often to investigate flowers or wet sand. They are occasionally attracted to over-ripe fruits and tree sap. Females mimic flight of their models and fly about slowly, seeking mates and their host plants on which to lay eggs. Males fond of hill-topping, patrolling a tree or rocky outcrop for hours with males of several other species.

Male underside

Yellow Kaiser

Male upperside

Freak ■ *Calinaga buddha* 9–10.5cm

DESCRIPTION Sexes similar. Relatively very short antennae and rufous thorax immediately distinguish this butterfly from all others. It varies along its range, with paler forms in western Himalaya. **DISTRIBUTION** Pakistan to North-east India. Very rare in west, but eastern populations appear to be better known. **HABITATS AND HABITS** Occurs from low elevation to more than 2,000m in western Himalaya. Both sexes attracted to over-ripe fruits and dung, with males also visiting wet sand. They form part of large congregations of butterflies at wet sand, but themselves rarely occur in any numbers.

Underside

Male upperside

Common Nawab ■ *Polyura athamas* 6–7.5cm

DESCRIPTION Sexes similar. Width of pale green band seasonally and individually variable. Only one pale green spot below forewing apex, which may be absent (red circle in inset). Width of pale green band across both wings very variable. Pallid Nawab (see p. 106) distinguished by paler greenish band across wings. **Anomalous Nawab** *P. agraria* (6–7.5cm; found with Common and shares same distribution and habits), distinguished by usually having two pale spots below forewing apex. In other features, as variable as Common. **DISTRIBUTION** Sri Lanka; peninsular India; Himalaya from Himachal Pradesh to North-east India, east to the Philippines and southwards to Indonesia. **HABITATS AND HABITS** A forest insect usually encountered sailing with powerful flight about the canopy, where males take up a beat and harry passing butterflies. Males attracted to wet sand. Both sexes fond of over-ripe fruits, tree sap, carrion and animal dung.

Male upperside

Male upperside

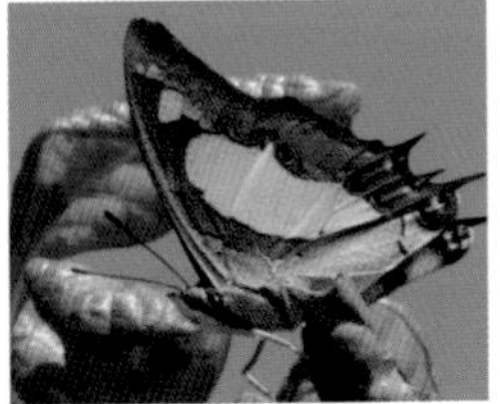

Underside Anomalous

Underside

Male upperside

Pallid Nawab ■ *Polyura arja* 7.5–8.5cm

DESCRIPTION Sexes similar. Very variable seasonally and individually, but can be distinguished from Common Nawab (see p. 105), which it closely resembles, by lack of any yellow shade on pale band across both wings. Shade is pale greenish-white, hence the name Pallid Nawab. **Malayan Nawab** *P. moori* (8–8.5cm; Sikkim to North-east India and Indonesia) has no dark shade at base of upperside hindwing; pale band not bounded by straight black borders, and outer border on forewing bulges into pale area above inner margin. No distinction on underside between Malayan and other similar members of the genus, except irregular shape of lower half of forewing pale band. **DISTRIBUTION** Along Himalaya from eastern Nepal to North-east India, Thailand and Vietnam. **HABITATS AND HABITS** Appears restricted to low elevation, where it occurs with other members of the genus. Males come to wet sand. Both sexes fond of over-ripe fruits, dung, rotting carcasses and tree sap. Males take up a beat and patrol it aggressively. In flight, chalky-white colour of pale areas is noticeable, and an experienced eye can separate this species from other members of the genus.

Underside

Malayan

Great Nawab ■ *Polyura eudamippus* 10–12cm

DESCRIPTION Sexes similar. Upperside black forewing border has two rows of pale spots. On underside, two vertical brown bands across forewing are never connected to form an 'H', as in **China Nawab** *P. narcaea* (7–7.5cm; Bhutan to China and Vietnam), which is smaller and has one row of pale spots on dark margin to upperside forewing. **DISTRIBUTION** Uttarakhand to North-east India, China and Thailand. **HABITATS AND HABITS** An inhabitant of forested hills at low elevation. Flight swift and powerful, consisting of a few flaps followed by a glide. Both sexes fond of over-ripe fruits, tree sap, dung and animal carcasses; males visit damp sand. Aggressive males can be attracted by decoys. They often settle in damp grass in between bouts of chasing off other males from their beat around a tree.

Male upperside

Underside

Underside China

Black Rajah ■ *Charaxes solon* 7–8cm

DESCRIPTION Sexes similar. Upperside and underside pattern singular in India. Vaguely similar **Blue Nawab** *Polyura schreiber* (9–10cm; along Western Ghats from Maharashtra to Kerala; Andaman Islands and from North-east India to the Philippines and Indonesia) is larger with blue band on upperside and rather different pattern on underside. **DISTRIBUTION** Throughout plains of India and Sri Lanka, except arid regions and at low elevation along Himalaya from Himachal Pradesh to North-east India, to the Philippines and Sulawesi. **HABITATS AND HABITS** Males come to wet sand. Both sexes often perch on high trees and descend readily to feed on over-ripe fruits, dung, carrion and tree sap.

Upperside

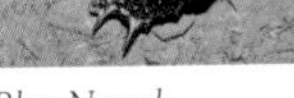

Blue Nawab

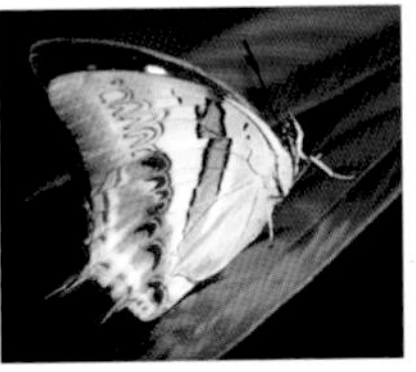

Blue Nawab

Underside

Plain Rajah ■ *Charaxes psaphon* 9–11.2cm

DESCRIPTION Male's upperside tawny with plain black border; female larger, with white band across forewing and tail on hindwing. In both sexes, upperside forewing dark border unmarked except in dry-season form, in which some tawny marks may be present in lower half of black border. In **Tawny Rajah** C. *bernardus* (Uttarakhand to the Philippines in similar localities), female has one form similar to illustrated male form *hindia*, but with tail on hindwing. Male has several forms, mainly varying in presence or absence of white band across forewing, and dark band on hindwing that may be continuous or broken up into spots – distinguished from all other species by broad dark border to forewing, which has a few tawny or pale spots on lower half. **DISTRIBUTION** Sri Lanka, South India to Odisha and North-east India. **HABITATS AND HABITS** Inhabits dense forests at low elevation, usually at foot of hills, ascending ranges to around 1,000m. Both sexes attracted to over-ripe fruits, tree sap, carcasses and animal dung. Males are pugnacious and attack passing butterflies from a prominent perch.

Male upperside

Male Tawny

Male Tawny hindia

Female

Variegated Rajah ■ *Charaxes kahruba* 9–12cm

DESCRIPTION Female larger than male and has a better developed tail on hindwing. Both sexes otherwise similar. Irregular dark band across undersides of both wings distinctive; on hindwing outer part of upper third of dark band is highly concave. On upperside, distinguished from Scarce Tawny Rajah (see below) and **Tawny Rajah** *C. bernardus* by yellow spots on dark forewing margin, reaching all the way to near apex. Distinguished from **Yellow Rajah** *C. marmax* (9–12cm; central Nepal to Vietnam) on upperside by line bordering inner edges of spots being much darker, and on underside by plain yellowish tint of Yellow, while Variegated is heavily marked; outer part of dark band much less concave in Yellow, and highly concave in Variegated. **DISTRIBUTION** At low elevation along Himalaya from Uttarakhand to North-east India, to China and Vietnam. **HABITATS AND HABITS** Flight powerful and rapid, and generally high around trees. Males are territorial. Males attracted to wet sand, but females rarely seen. Both sexes fond of over-ripe fruits, rotting carcasses, dung and tree sap.

Male upperside

Female Yellow

Male

Female upperside Yellow

Male Yellow

Scarce Tawny Rajah ■ *Charaxes aristogiton* 7–9.5cm

DESCRIPTION Sexes similar. Female has longer hindwing tail than male. On upperside, ochreous spots on dark forewing margin never reach space 7 (green arrow), leaving apical part of dark band broader than in Yellow and Variegated Rajahs (see above.) On underside, ground colour is more uniform, brown washed over with purple tint. **DISTRIBUTION** Central Nepal to north Vietnam. **HABITATS AND HABITS** A rather little-known species that probably has the same habits and distribution as other members of the genus. Confined to low elevation along Himalaya, with males recorded at wet sand. Both sexes fond of over-ripe fruits, dung, rotting carcasses and tree sap.

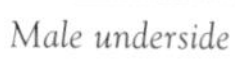
Male underside

Male upperside

Common Map ■ *Cyrestis thyodamas* 5–6cm

DESCRIPTION Sexes similar. Females of western Himalayan subspecies *ganescha* yellowish; males identical to South Indian subspecies *indica*. Wings angular with 'false head' painted on at bottom of hindwing. Similar **Marbled Map** C. *cocles* (5–7cm; Odisha to North-east India and Borneo) has three forms: pale (*cocles*), dark interspersed with white patches (*earlei*), and dark with central white band (*natta*). **DISTRIBUTION** Peninsular India and Himalaya to North-east India, Taiwan and Okinawa. **HABITATS AND HABITS** Occurs in humid forests, ascending to around 2,400m in Himalaya. Flight swift, consisting of a few flaps followed by a glide. Wings held slightly below horizontal when gliding. Males sometimes territorial. Both sexes fond of flowers and over-ripe fruits, and sometimes found at wet sand. The species appears to be distasteful to birds, yet individuals with 'false head' missing due to attack are not uncommonly seen.

Male underside

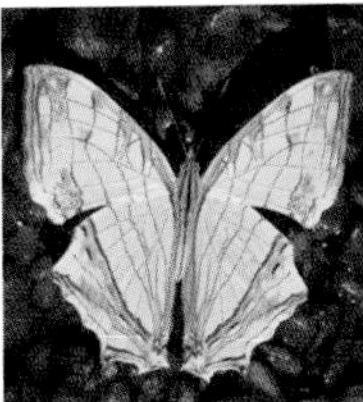

Male upperside

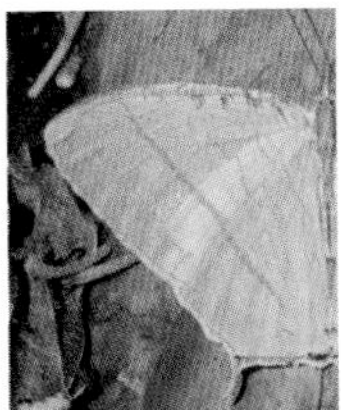

Marbled *cocles*

earlei

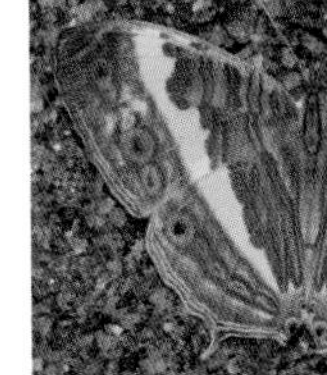

natta

Common Maplet ■ *Chersonesia risa* 4–4.5cm

DESCRIPTION Seasonal forms differ. On forewing, sixth line from base is straight (yellow arrow below) in this species, waved in **Wavy Maplet** C. *intermedia* (3.5–4.5cm; Assam to Vietnam). **DISTRIBUTION** Along Himalaya west of Uttarakhand to North-east India, onwards to Malaysia and Borneo at low elevation. **HABITATS AND HABITS** Prefers humid jungle up to 1,600m. Settles on undersides of leaves when disturbed. Fond of flowers and wet sand.

Male upperside

Upperside Wavy

Upperside Wavy

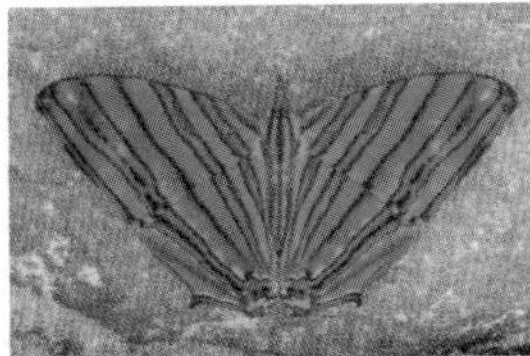

Male upperside

Tabby ■ *Pseudergolis wedah* 5.5–6.5cm

DESCRIPTION Sexes similar. A singular species, with some individual variation in shade of brown ground colour. **DISTRIBUTION** Along Himalaya from Himachal Pradesh to North-east India, eastwards to southern China and northern Vietnam. **HABITATS AND HABITS** Ascends to 2,500m in west. Rarely found away from forests. Commonly occurs along streams, where males take up a beat and confront passing butterflies in mid-air. Males sometimes visit wet sand, but are never found in any numbers. Both sexes fond of over-ripe fruits, but rarely visit other rotting substances.

Upperside

Underside

Male upperside

Popinjay ■ *Stibochiona nicea* 6–8cm

DESCRIPTION Sexes similar, but females duller than males. No similar species in India. **DISTRIBUTION** Himachal Pradesh to North-east India, China and Malaysia. **HABITATS AND HABITS** A resident of humid forest, found from low elevation to more than 2,000m in hills. Common in cardamom plantations in Sikkim. Flight a rapid flap and glide. Butterfly rarely closes its wings, preferring to hide flattened against underside of large leaf when disturbed. Males are pugnacious, but not quite as much as the rajahs. Both sexes fond of over-ripe fruits and tree sap, but do not seem to visit flowers.

Underside

Female upperside

Constable ■ *Dichorragia nesimachus* 6.5–8.5cm

DESCRIPTION Sexes similar. Singular pattern makes this butterfly impossible to confuse with anything else found in region. **DISTRIBUTION** Along Himalaya from Himachal Pradesh to North-east India, Japan and the Philippines. **HABITATS AND HABITS** In west, ascends to 1,800m, but most common at low elevation in North-east India. Butterfly of dense forests that ventures into clearings. Flight fast, using a flap-and-glide technique. Males often territorial, when they are just as pugnacious as skippers and rajahs. Males visit wet sand. Both sexes fond of over-ripe fruits and tree sap.

Underside

Male upperside

Common Jester ■ *Symbrenthia lilaea* 4.5–5.5cm

DESCRIPTION Sexes similar. Underside, coupled with shape of wings, is singular. Width of orange bands on upperside varies according to season. From above, it is nearly impossible to distinguish from other members of the genus. **Bluetail Jester** *S. niphanda* (Jammu and Kashmir to North-east India and Laos) is found at slightly higher elevation, generally at 1,500–2,600m. Above, both this and Common have a narrow orange streak below apex (vertical blue arrow); orange spot below this (horizontal blue arrow) often separate in Common, but always joined to subapical orange band in Bluetail. On underside, Bluetail can be distinguished from Himalayan Jester (see p. 112) and **Spotted Jester** *S. hypselis* by short black streak arising from base of forewing (horizontal green arrow) and running along top of cell for a short distance. On hindwing, one of the metallic greenish-centred crescents is missing (vertical green arrow). **DISTRIBUTION** Telangana to Odisha; Himachal Pradesh to North-east India, to Taiwan and Sulawesi. **HABITATS AND HABITS** Ascends hills to more than 2,000m. Flight swift, using a flap-and-glide technique like that of the sailers (*Neptis* spp.), but much more powerful. Males are territorial. Males come to wet sand. Both sexes fond of flowers.

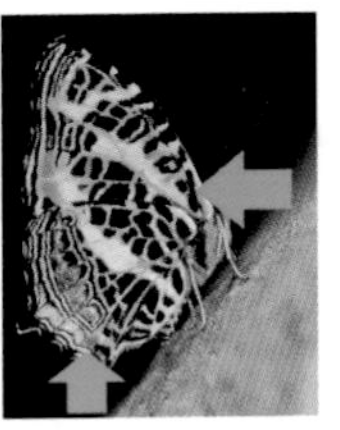
Bluetail

Upperside Bluetail

Upperside

Underside

Himalayan Jester ■ *Symbrenthia brabira* 4–5.5cm

DESCRIPTION Sexes similar. Both this and the **Spotted Jester** *S. hypselis* lack orange streak from upperside forewing apex that distinguish Common and Bluetail Jesters (see p. 111). Hindwing lacks all shining blue markings, except along edge of wing near tail, where there is a series of crescents with metallic greenish centres (horizontal green arrow). Spotted has row of large, blue-centred marks on underside hindwing; these markings have no metallic green scales in them in Himalayan (vertical green arrow and curved green line); instead of greenish-centred crescents near hindwing tail, Spotted has continuous band that has metallic greenish scales in it. **DISTRIBUTION** Jammu and Kashmir to North-east India and Taiwan. **HABITATS AND HABITS** An inhabitant of dense broadleaved forest at 1,500–2,750m in Himalaya. Generally found along forest streams, where males patrol a beat. Flight is powerful, using flap-and-glide technique. Both sexes visit flowers, and males come to wet sand.

Upperside Spotted

Underside Spotted

Upperside Spotted

Underside

Mongol ■ *Araschnia prorsoides* 5–5.5cm

DESCRIPTION Sexes similar. There is no similar butterfly in the area. **DISTRIBUTION** Nagaland and Manipur to northern Myanmar and western China. **HABITATS AND HABITS** Local species on the wing in summer. A forest insect. Flight a powerful flap and glide, generally high about trees, but males descend to wet sand. Both sexes visit flowers. Males are territorial, generally choosing a site on a tree to launch attacks on passing butterflies.

Upperside

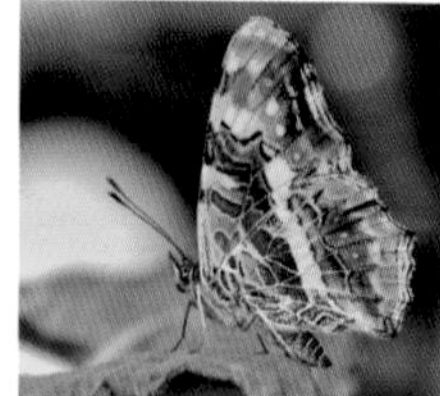

Underside

Indian Tortoiseshell ■ *Aglais cashmirensis* 5.5–6.5cm

Male upperside

DESCRIPTION Shape of wings and upperside pattern is singular below 3,000m in main Himalayan range and throughout outer ranges. Above that, two similar species occur, **Ladakh Tortoiseshell** *A. ladakensis* and **Mountain Tortoiseshell** *A. rizana*. **Large Tortoiseshell** *Nymphalis xanthomelas* (6–7cm; Europe to Japan and western Himalaya) is larger with orange ground colour. **DISTRIBUTION** Safed Koh in Afghanistan to Himalaya from Kashmir to Nagaland. **HABITATS AND HABITS** A hill species occuring from 400m to well over 3,000m. Very common in suitable localities. Generally a creature of open areas bordering forests, frequently visiting gardens and fields in rural landscapes. Several females lay their eggs together, so that when all the larvae emerge, they eat considerable amounts of their larval host plants, nettles. Males are territorial and rarely visit wet sand. Both sexes fond of flowers.

Underside

Large

Blue Admiral ■ *Kaniska canace* 6–7.5cm

DESCRIPTION Sexes similar. The only butterfly with non-iridescent blue band across wings. Himalayan subspecies *canace* has moderate blue band; southern Indian subspecies *viridis* has narrow greeenish-blue band on forewing; Sri Lankan subspecies *haronica* has broad blue band on forewing. **DISTRIBUTION** Hills of southern India from Karnataka to Kerala; Sri Lanka; Himalaya from Pakistan to North-east India, Siberia, Japan, the Philippines and Indonesia. **HABITATS AND HABITS** A forest insect. Males come occasionally to wet sand and are strongly territorial, staking out a beat along a forest stream and harrying passing insects from a few metres above the ground. Fond of over-ripe fruit and tree sap.

Upperside viridis

Underside viridis

Upperside haronica

Upperside canace

Painted Lady ■ *Vanessa cardui* 5.5–7cm

DESCRIPTION Sexes similar. Orange markings on upperside hindwing distinguish this species from Indian Red Admiral (see below). **DISTRIBUTION** Throughout Pakistan, India and Sri Lanka, to Europe, Africa, North America and Asia. **HABITATS AND HABITS** Flight swift and powerful, often crossing the Mediterranean and other seas during migrations. Both sexes fond of flowers. One of the most widespread butterflies.

Upperside

Underside

Indian Red Admiral ■ *Vanessa indica* 5.5–6.5cm

DESCRIPTION Sexes similar. Upper hindwing brown with black-spotted red border. Rather similar **Red Admiral** *V. atalanta* (5.5–6.5cm; North America, northern Africa, Europe to Balochistan) occurs in western Pakistan (Zhob, Sheen Ghar) and is distinguished by narrower red band across upperside forewing. **DISTRIBUTION** Sri Lanka; hills of Karnataka and Kerala, and along Himalaya from Pakistan to North-east India, Myanmar and Taiwan. **HABITATS AND HABITS** Occurs from around 400m to well over 2,500m. Forest insect that also ventures into gardens. Flight powerful and swift. Highly territorial and always found singly, though as soon as a beat is vacated, another individual appears within minutes to occupy the territory, which is usually at the level of bushes in sunny glades. Males rarely visit wet sand. Both sexes fond of flowers.

Upperside

Underside

Upperside Red Admiral

Peacock Pansy ■ *Junonia almana* 6–6.5cm

DESCRIPTION Sexes similar. Upperside is singular. Wet-season form underside has eyespots; dry-season form has angular wings, and underside resembles a dry leaf. **DISTRIBUTION** Throughout Pakistan, India and Sri Lanka to the Philippines and Sulawesi. **HABITATS AND HABITS** Common in sunny, open places near forests. Generally found at low elevation, but stragglers ascend hills to more than 2,000m. Flight swift, consisting of a flap and glide. Usually occurs singly, though males do not appear to be strongly territorial. Both sexes fond of flowers.

Underside wet season

Underside dry season

Upperside wet season

Grey Pansy ■ *Junonia atlites* 5.5–6.5cm

DESCRIPTION Sexes similar. Somewhat like other members of the genus, but pale grey ground colour is singular. **DISTRIBUTION** Wetter parts of eastern Pakistan, throughout India and Sri Lanka except arid regions, to Sulawesi. **HABITATS AND HABITS** Ascends Himalaya to 1,500m. Found in sunny, open places, especially on shores of waterbodies, since its host plant is semi-aquatic. Flight swift but not very powerful, consisting of a flap-and-glide progression. Both sexes fond of flowers.

Upperside

Underside

Underside

Blue Pansy ■ *Junonia orithya* 4–6cm

DESCRIPTION Sexes similar, but female has duller colours and larger red eyespots on hindwing. Upperside cannot be confused with that of any other species. Underside distinguished from that of Yellow Pansy (see below) by two reddish bars across forewing cell. **DISTRIBUTION** Tropical Africa to Pakistan, India and Sri Lanka, eastwards to Australia. **HABITATS AND HABITS** Common in sunny, open places. Flight rapid, close to the ground. Males are territorial, often patrolling a patch of hot, sunny road for hours at a time. Males may visit water. Both sexes fond of flowers.

Male upperside

Female upperside

Yellow Pansy ■ *Junonia hierta* 4.5–6cm

DESCRIPTION Sexes similar, but females have duller colours than males. Upperside singular. Underside differs from Blue Pansy's (see above) in forewing being yellow, without white band and reddish bars of Blue. **DISTRIBUTION** Throughout Africa to India and Sri Lanka, eastwards to China and Cambodia. **HABITATS AND HABITS** Found in Himalaya to more than 2,000m, but it is unlikely that these individuals breed there. Common in sunny, open places. Flight rapid, near the ground, consisting of a brisk flap and glide. Both sexes fond of flowers.

Female upperside

Male upperside

Underside

Lemon Pansy ■ *Junonia lemonias* 4.5–6cm

DESCRIPTION Sexes similar. Upperside pattern is singular. Underside ground colour varies from brown to pink. Wet-season form usually more heavily marked than dry-season form. **DISTRIBUTION** Pakistan; throughout India and Sri Lanka to Taiwan. **HABITATS AND HABITS** Frequently seen in open, sunny places. Commonly found basking in gardens and on roadsides. Flight swift, generally near the ground. Males rarely visit wet sand. Both sexes fond of flowers.

Upperside dry season

Underside

Male upperside wet season

Chocolate Pansy ■ *Junonia iphita* 5.5–8cm

DESCRIPTION Sexes similar. Some individuals have white spot at top of underside hindwing (form *siccata*). Shape of wings and colour are distinctive. Wet-season form darker than dry-season form. **DISTRIBUTION** Sri Lanka; from Pakistan to Borneo. **HABITATS AND HABITS** Generally found at low elevation, but ascends to Himalaya to more than 2,400m. More of a forest species than other members of the genus, rarely being found in open country. Both sexes fond of flowers, and sometimes attracted to animal dung.

Male upperside

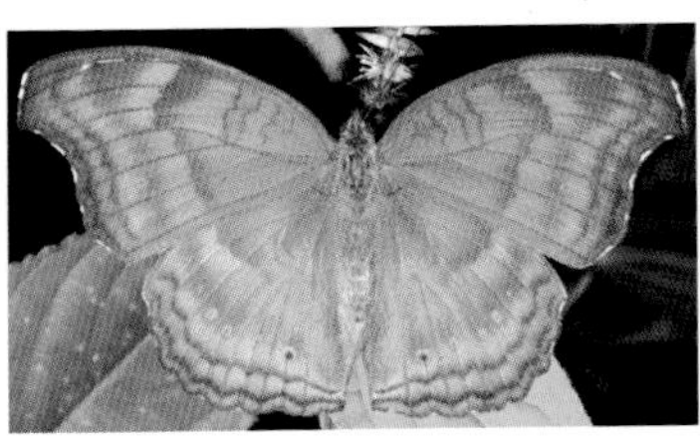

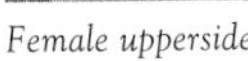

Female upperside

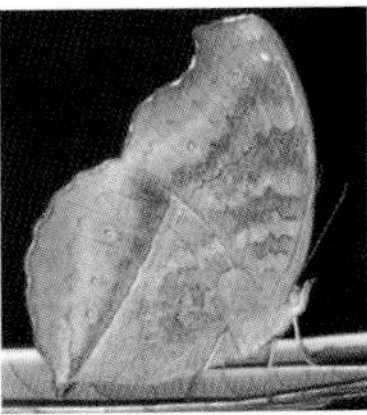

Underside iphita

Underside siccata

Great Eggfly ■ *Hypolimnas bolina* 7–11cm

DESCRIPTION Male has bluish-white patches on upperside; female mimics Common Crow (see p. 124), but wing shape is distinctive. Underside plain, dull brown in dry-season form, with broad white bands in wet-season form. Males never have light rufous-brown ground colour on underside, which is characteristic of Danaid Eggfly males (see below). **DISTRIBUTION** Madagascar; Arabia; throughout India and Sri Lanka to Australia. **HABITATS AND HABITS** Flight powerful and insect is capable of rapid progress. Females at times wonderfully mimic leisurely flight of their poisonous models, the brown crows (*Euploea* spp.). Both sexes fond of flowers.

Underside dry season

Male upperside

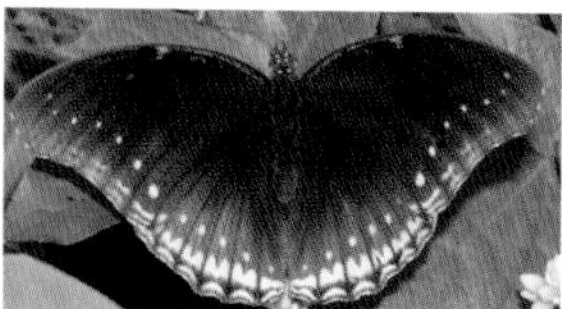
Female upperside

Underside wet season

Danaid Eggfly ■ *Hypolimnas misippus* 7–8.5cm

DESCRIPTION Male's underside bright brown with broad white bands; in female, single large black spot on upperside hindwing costa and crenulate hindwing border (blue circle in figure below) distinguish it from Plain Tiger (see p. 121). Form *inaria* mimics Plain Tiger form *dorippus*, lacking white band across forewing, while form *alcippoides* has white hindwings and mimics form *alcippoides* of Plain Tiger. **DISTRIBUTION** Northern South America; southern USA; Africa, throughout Pakistan, India and Sri Lanka to Taiwan and Australia.. **HABITATS AND HABITS** A forest insect, often occupying glades and open areas in the vicinity. Flight more powerful than Plain Tiger's. Males are territorial, often patrolling a beat along a stream or road. Both sexes fond of flowers.

Female upperside

Female underside

Male underside

Male upperside

Southern Blue Oakleaf ■ *Kallima horsfieldi* 8.5–11cm

Sri Lankan

DESCRIPTION Sexes similar. Pale blue band across forewing singular. **Sri Lankan Oakleaf** *K. philarchus* (8.5–11cm; endemic to Sri Lanka) and **Scarce Blue Oakleaf** *K. alompra* (9.5–11cm; Sikkim to southern Myanmar) almost identical, but easily distinguished by being the only such butterflies in their respective areas. Underside pattern highly variable, both individually and seasonally, identical to next species. **DISTRIBUTION** Western Ghats from Gujarat southwards to Kerala. **HABITATS AND HABITS** An insect of dense broadleaved forests at low elevation. Flight powerful and jerky, usually within shady forests. Butterfly settles frequently, though males sometimes patrol small beats in clearings or along paths, boisterously challenging passing butterflies. Fond of over-ripe fruits and tree sap.

Upperside Southern Blue

Upperside Scarce Blue

Orange Oakleaf ■ *Kallima inachus* 8.5–11cm

DESCRIPTION Sexes similar. Orange band across upperside forewing distinctive. Underside highly variable, individually and seasonally. Dry-season form has tip of forewing produced to a point. **White Oakleaf** *K. albofasciata* (same size; endemic to Andaman Islands) has whitish, rather than orange, band across forewing. **DISTRIBUTION** Himalaya from Pakistan to North-east India; Central Indian hills to Gujarat; Maharashtra; eastwards to Japan. **HABITATS AND HABITS** Ascends Himalaya to nearly 3,000m. Forest insect, never found in open country. Both sexes fond of over-ripe fruits and tree sap.

Upperside

Underside wet season

Underside dry season

Autumn Leaf ■ *Doleschallia bisaltide* 7.5–8.5cm

DESCRIPTION Sexes similar. Upperside pattern is singular. Underside of wet-season form darker than dry-season form's. **DISTRIBUTION** Sri Lanka; Western Ghats south from Maharashtra. Himalaya eastwards from central Nepal to China and Australia. **HABITATS AND HABITS** An inhabitant of dense forests, ascending Himalaya to around 1,800m. Flight boisterous and jerky, usually within forest or along streams and paths. Males come to wet sand. Both sexes fond of over-ripe fruits and flowers.

Underside

Upperside

Underside

Wizard ■ *Rhinopalpa polynice* 7–8cm

DESCRIPTION Sexes similar. Unmarked reddish-brown upperside with black border of variable thickness is singular. Underside similar to Blue Admiral's (see p. 113), but upperside immediately distinguishes this species. **DISTRIBUTION** Arunachal Pradesh to the Philippines and Sulawesi. **HABITATS AND HABITS** An inhabitant of dense forests at low elevation. Usually found within forests, but ventures into glades and along paths and streams. Settles frequently to bask. Both sexes attracted to over-ripe fruits and sap.

Upperside

Underside

Plain Tiger ■ *Danaus chrysippus* 7–8cm

DESCRIPTION Sexes similar, but male has additional black spot on hindwing. Hindwing margin not crenulate. Form *dorippus* lacks white band on forewing, and in form *alcippoides* all or most of hindwing is white. **DISTRIBUTION** Throughout India and Sri Lanka; Africa to southern Europe and Australia. **HABITATS AND HABITS** Flight slow, in keeping with its reputation for being highly poisonous. Both sexes fond of flowers. Plain Tiger is mimicked by several butterflies edible by birds, one of which – the Danaid Eggfly (see p. 118) female – even copies the forms of this species.

Underside

Upperside alcippoides

Male upperside

Underside dorippus

dorippus

Common Tiger ■ *Danaus genutia* 7.5–9.5cm

DESCRIPTION Sexes similar, but male has white-centred black patch on underside hindwing. Dark lines across orange area of both wings are singular. **White Tiger** *D. melanippus* (8–9.5cm; Odisha to Sulawesi) lacks most of orange on hindwing, but has more or less the same markings. **DISTRIBUTION** Afghanistan to Pakistan, throughout India and Sri Lanka to Australia. **HABITATS AND HABITS** Migrant species, with large congregations overwintering in some southern Indian valleys. However, they do not seem to return to the same spot. Flight weak, a few metres above the ground. Males congregate around plants containing pyrrolizidine alkaloids, such as *Crotalaria* spp. Their bodies contain poison. Both sexes fond of flowers.

Male underside

Female underside

Upperside White

Underside White

Female upperside

Glassy Tiger ■ *Parantica aglea* 7–8.5cm

DESCRIPTION Sexes similar, but male has a black patch near bottom of hindwing on both surfaces (horizontal red arrow). Ground colour pale grey. Distinctive pale lines on forewing (vertical red arrow) are elongate in this species. **DISTRIBUTION** Sri Lanka to Gujarat and West Bengal. Humid forests in peninsular India; throughout Himalaya east of Jammu and Kashmir to North-east India, Taiwan and Thailand. **HABITATS AND HABITS** Flight slow, along a forest path or in a clearing, where butterflies circle for hours, never attacking another butterfly but attracting a female by releasing pheromones from a brush at the tip of the abdomen. Males attracted to plants containing pyrrolizidine alkaloids, such as *Crotalaria* spp. Their bodies contain poison. Both sexes fond of flowers.

Male upperside

Female underside

Male underside

Male upperside

Sri Lankan Tiger

■ *Parantica taprobanis* 8.5–9.5cm

DESCRIPTION Sexes similar. Singular in Sri Lanka. Very similar **Nilgiri Tiger** *P. nilgiriensis* (8–9cm; hills of Tamil Nadu and Kerala above 1,500m) is darker on upperside, and inhabits shola forests; it ventures into nearby gardens and cultivation. **DISTRIBUTION** Endemic to hills of Sri Lanka. **HABITATS AND HABITS** Occurs above 1,000m. Flight slow and leisurely, with butterflies settling frequently on flowers. Usually found singly. Males attracted to plants containing pyrrolizidine alkaloids, such as *Ageratum* spp. Their bodies contain poison.

Underside

Underside Nilgiri

Chestnut Tiger ■ *Parantica sita* 8.5–11cm

DESCRIPTION Sexes similar, but males have prominent black patch near bottom of hindwing. Similar **Chocolate Tiger** *P. melaneus* (8.5–9.5cm; Uttarakhand to North-east India, northern Vietnam and Malaysia; 400–2,000m) is distinguished by being darker, with chocolate, not reddish-brown border to hindwings. **DISTRIBUTION** Jammu and Kashmir to North-east India, Japan, Ussuri and Thailand. **HABITATS AND HABITS** An inhabitant of dense, humid forest on hills, at 400–2,500m. Usually found sailing gracefully along forest paths or patrolling a glade. Females generally encountered within the forest. Both sexes fond of flowers and males come to wet sand. Males attracted to plants containing pyrrolizidine alkaloids, such as *Ageratum* spp. Their bodies contain poison.

Chocolate

Male upperside Chocolate

Male upperside

Male underside

Blue Tiger ■ *Tirumala limniace* 9–10cm

DESCRIPTION Sexes similar, but male has flap on underside of hindwing (red arrow). Ground colour blue, and blue streaks on forewing (circled in red) broader than in **Dark Blue Tiger** *T. septentrionis* (8–10.5cm; Sri Lanka to Gujarat and Odisha; Himachal Pradesh to North-east India, Taiwan and Indonesia). Hindwing cell has narrow dark streak, which is much broader in Dark Blue. **DISTRIBUTION** Afghanistan, throughout Pakistan, India and Sri Lanka to Taiwan and Sulawesi. **HABITATS AND HABITS** A strong-flying migrant that is found all over India. Breeds in forested areas. During migrations, this species, along with other members of the subfamily, roosts by the thousand on select trees and shrubs. Both sexes fond of flowers. Males attracted to plants containing pyrrolizidine alkaloids, such as *Crotalaria* spp. Their bodies contain poison.

Underside

Underside Dark Blue

Male upperside

Male upperside Dark Blue

Malabar Tree Nymph ■ *Idea malabarica* 11–16cm

DESCRIPTION Sexes similar, but female has broader wings than male. Singular in southern India. **Ceylon Tree Nymph** *I. iasonia* (12–14cm; Sri Lanka) and **Tavoy Tree Nymph** *I. agamarschana* (12–14cm; Andaman Islands and Myanmar) are very similar and singular in Sri Lanka and Andaman Islands respectively. **DISTRIBUTION** Western Ghats from Maharashtra southwards to Kerala. **HABITATS AND HABITS** Inhabits dense evergreen forests with heavy rainfall, where males take up a beat in clearings. Flight very slow, almost floating while patrolling or courting, but can be powerful when butterfly is travelling. Both sexes visit flowers. Males attracted to plants containing pyrrolizidine alkaloids. Their bodies contain poison.

Tavoy

Upperside

Ceylon

Common Crow ■ *Euploea core* 8.5–9.5cm

DESCRIPTION Sexes similar. On upperside, males lack two horizontal, parallel dark marks at bottom of upperside forewing, characteristic of southern Indian subspecies of Double-branded Crow males (see opposite). Sri Lankan subspecies *asela* has reduced white spots on forewing. On underside, Common can be distinguished from Double-branded by lack of two vertical rows of white spots on outer half of forewing (circled in red in Double-branded). **DISTRIBUTION** Pakistan, throughout India and Sri Lanka to Australia. **HABITATS AND HABITS** Ascends hills to more than 2,000m. Migrant in parts of India. Flight slow and leisurely, but capable of being sustained over long distances. Both sexes fond of flowers. Males congregate on plants containing pyrrolizidine alkaloids, such as *Crotalaria* spp. In evergreen forests in southern India, thousands may gather in congregations under shady banks by roadsides. Their bodies contain poison.

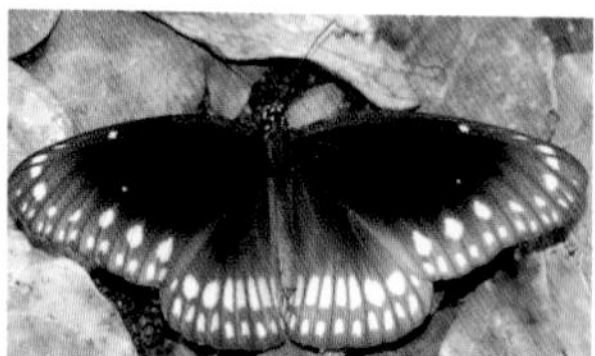

Subspecies core

Subspecies core

Subspecies asela

Double-branded Crow ■ *Euploea sylvester* 8.5–10.5cm

DESCRIPTION Sexes similar, but males have two prominent parallel brands along bottom of upperside forewing. Southern Indian subspecies *coreta* and Sri Lankan subspecies *montana* are brown on upperside, while North-east Indian subspecies *hopei*, and Andaman Islands and Myanmar subspecies *harrisi*, are iridescent blue. On underside forewing, species is distinguished by presence of three white spots (red arrow), and large white spot at end of forewing cell. Common Crow (see opposite) lacks these, except for cell spot. **DISTRIBUTION** From mountains of Sri Lanka to Maharashtra; eastern Nepal to Australia. **HABITATS AND HABITS** A butterfly of dense, humid forests, found alongside Common Crow in southern India. Migrant in southern India, but not in North-east India. Flight slow and leisurely, with both sexes being fond of flowers. Their bodies contain poison.

hopei

Male hopei *showing brands*

Male upperside coreta

Male coreta

King Crow ■ *Euploea klugii* 8.5–10cm

DESCRIPTION Sexes similar, but inner margin in female is straight, and convex in male. Wings much broader than those of either Common or Double-branded Crow (see opposite and above). On underside, species is easily distinguished by absence of white spot in forewing cell. Southern Indian subspecies *kollari* and Sri Lankan subspecies *sinhala* are brown, while North-east Indian subspecies *klugii* is iridescent blue. Both blue and brown forms occur together in North-east India. **DISTRIBUTION** Sri Lanka to Gujarat and West Bengal; Bihar, central Nepal to North-east India and Thailand. **HABITATS AND HABITS** Occurs in evergreen forest at low elevation. Often found in company of other crows. Both sexes fond of flowers, and males congregate on plants containing pyrrolizidine alkaloids, such as *Crotalaria* spp. Their bodies contain poison.

Underside kollari

Female upperside klugii

Male upperside kollari

Spotted Blue Crow ■ *Euploea midamus* 9.5–10.5cm

DESCRIPTION Sexes similar, but males have convex inner margin of forewing, while this is straight in females. Wide wings with white spot in forewing cell are distinctive and serve to distinguish this species from King Crow (see p. 125) at a glance. **DISTRIBUTION** Uttarakhand to North-east India, the Philippines and Indonesia. **HABITATS AND HABITS** An inhabitant of dense forest at low elevation, where males patrol small beats among bushes or in sunny glades. Both sexes fond of flowers, and males visit plants like *Crotalaria* spp. in order to obtain pyrrolizidine alkaloids for courtship. Their bodies contain poison.

Male upperside

Male underside

Male upperside

Striped Blue Crow ■ *Euploea mulciber* 9–10cm

DESCRIPTION In both sexes upperside forewing is iridescent blue, with spot in cell. Male hindwing unmarked, lower half brown, upper half greyish, with small yellow brand in cell; this distinguishes it from female of Double-branded Crow (see p. 125), in which entire hindwing is uniform brown. Female's hindwing striped with white. On underside, both sexes have spot in forewing cell. Male's hindwing has post-discal band composed of pairs of white spots (red arrow), while female has the same white stripes as on upperside. In Double-branded, this series is made up of one spot above each vein. **DISTRIBUTION** Telangana to West Bengal; Jammu and Kashmir to North-east India, the Philippines and Indonesia. **HABITATS AND HABITS** Occurs in dense forest in areas of heavy rainfall, ascending Himalaya to 2,000m. Usually found singly. Males sometimes patrol small beats along forest paths or streams, taking position about 5m above the ground. Both sexes fond of flowers, and males may visit wet sand. Males need to obtain pyrrolizidine alkaloids from plants like *Crotalaria* spp. for successful courtship. Their bodies contain poison.

Male upperside

Female upperside

Male underside

Long-branded Blue Crow ■ *Euploea algea* 9.5–10cm

DESCRIPTION Single white spot and long brand along bottom edge of upperside forewing distinguish male from female. The female lacks marginal spots on upperside, and upperside hindwing is similar to hindwing of female King Crow (see p. 125). **DISTRIBUTION** Eastern Nepal to Sulawesi. **HABITATS AND HABITS** Inhabits dense, humid forest at low elevation. Flight leisurely, with males patrolling a small beat in wait for females. Males come to wet sand, and both sexes visit flowers. Males require pyrrolizidine alkaloids from plants like *Crotalaria* spp. to complete successful courtship. Their bodies contain poison.

Male upperside

Male underside

Magpie Crow ■ *Euploea radamanthus* 8–9cm

DESCRIPTION Males are singular in subfamily Danainae and its mimics – female forms of Courtesan (see p. 103) and a female form of **Great Mime** *Papilio paradoxa* can be distinguished from them because neither of these mimics has the rounded forewings of Magpie Crow. In female, most of hindwing is white and paler than male's. **DISTRIBUTION** Central Nepal to Borneo. **HABITATS AND HABITS** Flight slow and leisurely, generally high about the canopy of forests, but both sexes descend readily to flowers. Males are fond of wet sand. Their bodies contain poison.

Male upperside

Male underside

Large Silverstripe ■ *Argynnis childreni* 7.5–10cm

DESCRIPTION Sexes similar. Large size and blue edging to upperside hindwing are distinctive. Underside has silver stripes, not spots. Similar but much smaller **Common Silverstripe** A. *kamala* (6.5–7.5cm; Pakistan to central Nepal) lacks blue on upperside hindwing, and silver stripe across underside hindwing is broken up into spots. **DISTRIBUTION** Himalaya from Himachal Pradesh to North-east India, Myanmar and China. **HABITATS AND HABITS** Occurs at 1,200–2,800m in glades in broadleaved forests. Flight powerful and jerky, usually along forest paths and on meadows adjoining forests. Both sexes fond of flowers.

Common Silverstripe male

Male upperside

Underside

Indian Fritillary ■ *Argynnis hyperbius* 6.5–8.5cm

DESCRIPTION Male has dark border to upperside hindwing; female is black spotted on upperside; underside pattern distinctive. Females from Palni Hills in Tamil Nadu lack white band across forewing and are duller than males (subspecies *castetsi*); subspecies *hybrida* (Nilgiri Hills) and *taprobana* (Sri Lanka) similar to North Indian subspecies *hyperbius*. **DISTRIBUTION** Sri Lanka; Abyssinia through Arabia to India, Indonesia and Australia. **HABITATS AND HABITS** Generally found in hills, especially in southern India, but also recorded from Chambal ravines in Madhya Pradesh. Never found in dense forests. Flight rapid and low over grassy areas. Females at times affect leisurely flight of their models, the Plain and Common Tigers (see p. 121). Both sexes fond of flowers.

Female castetsi

Female upperside hyperbius

Male upperside hyperbius

Underside taprobana

Queen of Spain Fritillary ■ *Issoria lathonia* 5–6cm

DESCRIPTION Sexes similar. Underside has extensive silver patches, which immediately distinguish this species. **DISTRIBUTION** Europe to Russia, along Himalaya above 1,800m to China. **HABITATS AND HABITS** Favours open, sunny meadows and ridges. Flies low over grass with flap-and-glide motion. Flight swift, with butterflies settling frequently to bask. Both sexes come readily to flowers.

Male upperside

Underside

Small Leopard ■ *Phalanta alcippe* 3.5–5cm

DESCRIPTION Sexes similar. Both sexes distinguished by black border on upperside wings, which Common Leopard (see p. 130) lacks. Subspecies *ceylonica* occurs in Sri Lanka, *mercea* in southern India, *alcippoides* in Himalaya and North-east India, *fraterna* in Nicobar Islands and *andamana* in Andaman Islands. **DISTRIBUTION** Sri Lanka; Maharashtra to Kerala; Sikkim to New Guinea. **HABITATS AND HABITS** A forest butterfly, found at low elevation in humid regions. Flight rapid, with continuous wingbeats. Usually occurs along edges of forests, where both sexes avidly feed on flowers. They sometimes visit dung and rotting matter. Males visit wet sand.

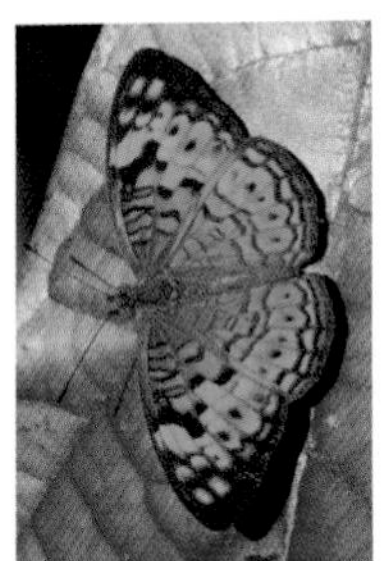

Upperside andamana

Underside alcippoides

Upperside mercea

Common Leopard ■ *Phalanta phalantha* 5–6cm

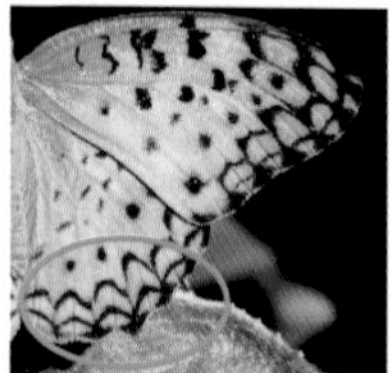

DESCRIPTION Sexes similar. Hindwing lacks dark border (blue circle). Relatively unmarked underside distinguishes it from silverstripes and fritillaries. **DISTRIBUTION** Sri Lanka; Pakistan, through India to Japan and Australia. **HABITATS AND HABITS** Larval host plants are species of willow, so butterfly is common near waterbodies at low elevation, ascending Himalaya to around 2,000m. Flight rapid, consisting of continuous series of wingbeats. Males are territorial. Both sexes fond of flowers.

Upperside

Underside

Rustic ■ *Cupha erymanthis* 5–6cm

Underside

DESCRIPTION Sexes similar. Broad, creamy band on upper forewing is distinctive. Underside pattern variable, pale or dark, depending on season. **DISTRIBUTION** Sri Lanka; Maharashtra south to Kerala; Himalaya from Uttarakhand eastwards, at low elevation. **HABITATS AND HABITS** An inhabitant of forests in areas of heavy rainfall. Flight hurried, jerky, usually along edges of forests, but often venturing into forest. Males occasionally territorial. Both sexes fond of flowers.

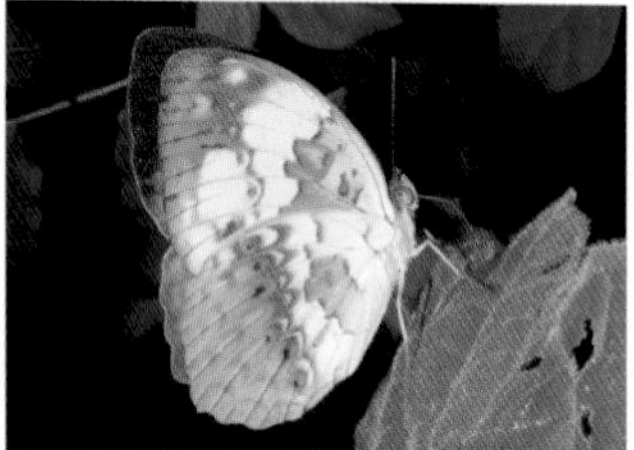

Underside

Upperside

Vagrant ■ *Vagrans egista* 5.5–6.5cm

DESCRIPTION Sexes similar. Wing shape and pattern are singular. Somewhat similar male Cruiser (see below) is much larger. **DISTRIBUTION** Odisha, Jharkhand; along Himalaya from Himachal Pradesh to North-east India, south to Australia. **HABITATS AND HABITS** Flight rapid and jerky, consisting of continuous wingbeats. Generally stays rather high, and males take up a beat several metres above the ground around trees. Males attracted to wet sand. Both sexes fond of flowers.

Upperside

Underside

Cruiser ■ *Vindula erota* 9–11cm

DESCRIPTION Large size, tail on hindwing and pattern are distinctive. Outer half of upperside wings of males of southern Indian subspecies *saloma* is pale. There is seasonal variation in female of Himalayan subspecies *erota*. **DISTRIBUTION** Sri Lanka, southern India north to Maharashtra, Himalaya from Nepal to North-east India and southwards to Sulawesi. **HABITATS AND HABITS** A forest insect. Flight swift, gliding and followed by a few flaps of wings. Males come to wet sand. Both sexes fond of flowers.

Male upperside

Female upperside wet season

Female upperside dry season

Female underside

Large Yeoman ■ *Cirrochroa aoris* 8–9cm

DESCRIPTION Male's upperside orange; female's duller and suffused with dark scales, except for narrow central pale band with jagged edges. Distinguished from **Common Yeoman** *C. tyche* (6.5–8.5cm; Nepal to the Philippines) by forewing tips, which are somewhat pronounced in this species and plain in Common. Sexes are quite similar in Common. **DISTRIBUTION** Nepal to North-east India and southern Myanmar. **HABITATS AND HABITS** A low-elevation butterfly, found in dense evergreen forests with high humidity. Flight strong, and males do not seem to be very territorial. Males attracted to wet sand and perspiration. Both sexes settle frequently on dung and flowers.

Common

Male upperside

Female upperside

Tamil Yeoman ■ *Cirrochroa thais* 6–7cm

Underside lanka

DESCRIPTION Sexes similar, but females more heavily marked on upperside than males. Species is singular in southern India (subspecies *thais*) and Sri Lanka (subspecies *lanka*). Some individuals have white stripe across middle of wings on underside. **DISTRIBUTION** Endemic to Sri Lanka, and Western Ghats and hills of southern India from Gujarat to Kerala. **HABITATS AND HABITS** Inhabits dense forests at low elevation. Flight moderately powerful, and insect settles frequently. Females usually seen within forest. Males attracted to wet sand and perspiration.

Male lanka

Male thais

Common Castor *Ariadne merione* 4.5–6cm

Underside

DESCRIPTION Sexes similar. Wing margins not highly angled. Dark line on forewing not highly angled (see Angled Castor, below). Ground colour pale to dark brown, and in some West Himalayan individuals some bands are almost yellow, contrasting sharply with brown bands on wing. **DISTRIBUTION** Throughout Sri Lanka, and from Pakistan eastwards to the Philippines and Borneo. **HABITATS AND HABITS** Ascends Himalaya to 1,500m. Common wherever castor-oil plants grow, but usually avoids dense forests. Flight weak, with butterflies settling frequently to bask or visit flowers. Usually settles with wings open. Occasionally visits over-ripe fruits.

Upperside

Upperside

Angled Castor ■ *Ariadne ariadne* 4.5–6cm

Underside

DESCRIPTION Sexes similar. Forewing margins angled (red arrow), and hindwing margins are crenulate (blue arrow). Dark line across forewing highly angled in middle (green arrow). **DISTRIBUTION** Sri Lanka; India to Hainan and Sulawesi. **HABITATS AND HABITS** Generally confined to low elevation in humid areas, avoiding dense forests. Common wherever castor-oil plants grow, but rarely found in company of Common Castor (see above). Flight weak and sailing. Both sexes fond of flowers.

Female upperside

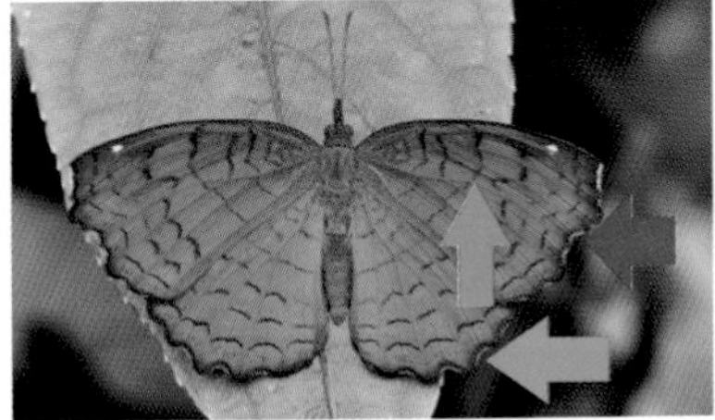
Male upperside

Joker ■ *Byblia ilithyia* 4.5–5.5cm

DESCRIPTION Sexes similar. Upperside markings are singular. Width of dark markings varies individually. **DISTRIBUTION** Peninsular India and Sri Lanka to Arabia and South Africa. **HABITATS AND HABITS** Locally common, in open areas where its larval food plant grows. Inhabits dry regions. Flight weak, and near the ground. Both sexes fond of flowers. One of several African butterflies that have colonized the subcontinent.

Underside

Male upperside

Female upperside

Common Sailer

■ *Neptis hylas* 5–6cm

DESCRIPTION Sexes similar. Ochreous brown shade on underside singular. On underside, white bands outlined with black lines. Wet-season form darkest, with narrow white bands. On Indian subcontinent, only **Pallas's Sailer** *N. sappho* (Europe to Japan; along Himalaya from Pakistan to North-east India) also has white bands on underside outlined with black; it is easily distinguished by darker brown, almost chocolate ground colour. On upperside, there is no distinguishing feature between the two species. **DISTRIBUTION** Sri Lanka; peninsular India; Himalaya and North-east India. **HABITATS AND HABITS** A forest insect that ventures into nearby open areas. Flight jerky and gliding. Males occasionally territorial. Both sexes fond of flowers.

Underside Pallas's

Upperside varmona

Underside varmona

Chestnut-streaked Sailer ■ *Neptis jumbah* 6–7cm

DESCRIPTION Sexes similar. Upperside hindwing has only one broad white band. Similar **Short-banded Sailer** *Phaedyma columella* (6.5–7.5cm; peninsular India to Gujarat; Uttarakhand to North-east India and Sulawesi) has two white bands on upperside hindwing. On underside, it lacks black spot in cell of hindwing that is prominent in Chestnut-streaked. **DISTRIBUTION** Endemic to Sri Lanka, India and Myanmar. In India, found almost throughout peninsula and again from Sikkim to North-east India. **HABITATS AND HABITS** A forest insect rarely venturing out into open country. Flies rather powerfully about gardens and along forest clearings. Flight typically the flap and glide after which this genus got its name. Both sexes fond of flowers and settle frequently to bask.

Upperside and underside Short-banded

Upperside

Underside

Common Lascar ■ *Pantoporia hordonia* 4.5–5cm

DESCRIPTION Sexes similar; smaller than most similar species. Underside pattern distinguishes members of this genus from sailers (*Neptis* spp.). Nearly identical **Extra Lascar** *P. sandaka* has large grey area on upper hindwing where forewings and hindwings overlap (shown below with a green line). Both species very variable, and this is the only feature that distinguishes them externally. Species occur together over most of their distribution except in Sri Lanka, where only Common has been recorded. **DISTRIBUTION** Sri Lanka; peninsular India; Himalaya from Uttarakhand eastwards to Borneo. **HABITATS AND HABITS** A forest insect, hardly ever found away from low-elevation broadleaved forest, with stragglers ascending Himalaya to 1,500m. Flight slow, within forest or along edges. Settles frequently; both sexes fond of flowers.

Upperside

Extra

Upperside

Underside

Orange Staff Sergeant ■ *Athyma cama* 6–7.5cm

DESCRIPTION Male smaller than female, with obscure red streak in upperside forewing cell and red spot below forewing apex. Female has white band across abdomen on upperside, and narrower, more sharply defined orange bands across both wings than female of Colour Sergeant (see below). **DISTRIBUTION** Along Himalaya from Uttarakhand to North-east India, to Taiwan and Borneo. **HABITATS AND HABITS** Ascends hills to at least 2,400m. A forest insect, venturing into gardens and glades, where it visits flowers. Males are territorial and aggressively attack passing butterflies.

Male underside

Male upperside

Female upperside

Male underside

Colour Sergeant ■ *Athyma nefte* 5.5–7cm

DESCRIPTION White, black and orange bands on hindwing of male are singular. In female, orange and dark bands on upperside are not sharply defined and are very broad, often merging into each other. **DISTRIBUTION** South India south of Karnataka; Uttarakhand to North-east India and Borneo. **HABITATS AND HABITS** Inhabits forested regions at low elevation. Flight powerful and swift, using a snappy flap-and-glide technique. Males often found at wet sand, while females rarely leave the forest. Both sexes fond of flowers.

Female upperside

Male upperside

Himalayan Sergeant ■ *Athyma opalina* 6–7cm

DESCRIPTION Sexes similar. Distinguished from **Oriental Sergeant** *A. orientalis* (5.5–7cm; Uttarkhand to North-east India) by broader, pure white bands on upperside, which are narrow and slightly suffused with dark scales in Oriental. Both Himalayan and Oriental Sergeants may be distinguished from Staff Sergeant (see p. 138) female by lack of black mark around centre of the three white cell spots on underside forewing, and faintly marked dark spots between two white bands on underside hindwing, which are always prominently marked in Staff. **DISTRIBUTION** Pakistan along Himalaya to North-east India and Taiwan. **HABITATS AND HABITS** The most frequently seen member of the genus in west Himalaya, where it inhabits broadleaved forest and scrubland at 1,000–2,800m. Common visitor to gardens in hills. Flight consists of the usual flap and glide common to the genus. Males territorial, generally staking out a territory that can be patrolled from a bush along a forest path or stream. Males visit wet sand. Both sexes fond of flowers and dung.

Oriental

Underside

Underside Oriental

Male upperside

Common Sergeant ■ *Athyma perius* 6–7cm

DESCRIPTION Sexes similar. Ochreous brown underside colour and line of black spots on hindwing white band are distinctive. Abdomen striped with white. **Studded Sergeant** *A. asura* (6–7.5cm; Himachal Pradesh to North-east India above 1,500m) is much darker brown on underside and has the row of spots on hindwing in middle of the white patches, not at upper end as in Common. **DISTRIBUTION** Sri Lanka, peninsular India, Himalaya from Himachal Pradesh to North-east India. **HABITATS AND HABITS** Flight consists of the usual flap and glide, much more powerful than in sailers (*Neptis* spp.). Males territorial, sometimes patrolling bush-height beats, and sometimes in the tree canopy. They may visit wet sand. Both sexes fond of flowers.

Upperside

Underside

Upperside Studded

Blackvein Sergeant ■ *Athyma ranga* 6–7cm

DESCRIPTION Sexes as illustrated; both sexes are singular among Indian butterflies. **DISTRIBUTION** Along Western Ghats from Karnataka to Kerala; central Nepal to Thailand. **HABITATS AND HABITS** Confined to dense, humid forests at low elevation. Flight swift, using flap-and-glide technique. Males visit wet sand, and both sexes are fond of flowers. They are sometimes attracted to dung and over-ripe fruits.

Male upperside

Female upperside

Male underside

Staff Sergeant ■ *Athyma selenophora* 5.5–7.5cm

DESCRIPTION Males similar to males of **Small Staff Sergeant** *A. zeroca*, but white band on forewing does not have regular edges; on underside, pale band in forewing cell is continuous in Small, and broken into four sections in Staff (red arrow). Female of Staff can be distinguished from Himalayan and Oriental Sergeants (see p. 137) and Small female by fourth white mark in underside forewing cell (red arrow) not being in line with marks below it, and dark marks on hindwing (red circle) being well developed. **DISTRIBUTION** Peninsular India as far north as Goa in west, and to North-east India in east; along Himalaya from Uttarakhand to Myanmar and Borneo. **HABITATS AND HABITS** Found in forests, ascending Himalaya to around 2,000m. Flight swift and powerful, using typical flap-and-glide technique preferred by the genus. Females visit wet sand with males. Both sexes fond of flowers, and rarely visit over-ripe fruits.

Male

Male small

Male

Male small

Female upperside

Female underside

Commander ■ *Moduza procris* 6–7.5cm

DESCRIPTION Sexes similar. Singular in peninsular India and Sri Lanka. In eastern Himalaya, **Scarce White Commodore** *Sumalia zulema* (8–8.5cm; eastern Nepal to North-east India and northern Vietnam) is similar, but lacks white mark in forewing cell, and white band across wings has straight edges. **DISTRIBUTION** Sri Lanka; peninsular India; Himalaya from Uttarakhand eastwards to North-east India, the Philippines and Indonesia. **HABITATS AND HABITS** Generally not found away from forests. Stragglers have been recorded in Himalaya as high as 1,500m. Flight powerful, comprising a flap and glide interspersed with rapid series of flaps. Males visit wet sand and are sometimes territorial, usually at level of bushes or small trees. Both sexes fond of flowers.

Scarce White Commodore

Upperside

Underside

Knight ■ *Lebadea martha* 5.5–6.5cm

DESCRIPTION White band across both wings is narrow, and tip of forewing apex is white. Narrow forewings are singular, the only similar species being Commander and Scarce White Commodore (see above). **DISTRIBUTION** Eastern Nepal eastwards to North-east India, China and Indonesia. **HABITATS AND HABITS** A forest insect. Restricted to low elevation along its range. Flight not very powerful and fluttering, but capable of bursts of speed. Flies in the undergrowth, often coming out to visit flowers. Males visit wet sand.

Upperside

Clipper ■ *Parthenos sylla* 9.5–13cm

DESCRIPTION Wing shape and markings distinctive. Sri Lankan subspecies *cyaneus* pale greyish-green; southern Indian subspecies *virens* golden-green, and North-east Indian *gambrisius* bluish-green. Andaman Islands subspecies *roepstorfii* has white forewing apex. Pale markings wider and forewing apex often tipped with white in dry-season form. **DISTRIBUTION** Sri Lanka, southern India to Goa; North-east India to the Philippines and New Guinea. **HABITATS AND HABITS** A forest insect, usually seen flying swiftly along forest streams and paths, or around glades. Flight swift and gliding, with wings held characteristically below the horizontal. Both sexes fond of flowers.

Upperside roepstorfii

Wet-season virens

Dry-season virens

Upperside gambrisius

Grey Count ■ *Cynitia lepidea* 6.5–8cm

DESCRIPTION Sexes similar, but margin of forewing less arched in female than in male. Species is singular – no other species has grey area of similar shape. **DISTRIBUTION** Kerala to Maharashtra on east, to North-east India on west; along Himalaya from Uttarakhand to Myanmar, Vietnam and Indonesia. **HABITATS AND HABITS** A jungle insect found at low elevation, though stragglers have been recorded up to 2,000m. Flight swift, with a powerful flap-and-glide technique. Males often seen at wet sand. Both sexes fond of over-ripe fruits and tree sap.

Male upperside

Female upperside

Plain Earl ■ *Tanaecia jahnu* 6–8cm

DESCRIPTION Sexes as illustrated. Distinguished from Common Earl (see below) by dark lines across forewing, composed of crescents in Plain, and dots in Common. **DISTRIBUTION** Sikkim to Malaysia. **HABITATS AND HABITS** Inhabits dense forest at low elevation in regions of heavy rainfall. Flight powerful and rapid. Males are territorial, generally selecting a beat along a forest path. Both sexes fond of over-ripe fruits and tree sap.

Male

Female

Common Earl ■ *Tanaecia julii* 6.5–8cm

DESCRIPTION Sexes similar, but most males have distinctive blue border to hindwing, which females lack. This border often reduced or lacking in males from Meghalaya. Distinguished from Plain Earl (see above) by dark lines across forewing being composed of spots, not crescents. Ground colour of underside is variable, from yellowish to bluish. **DISTRIBUTION** Uttarakhand eastwards along Himalaya to Hainan and Indonesia. **HABITATS AND HABITS** Inhabits dense forest at low elevation, with stragglers recorded as high as 1,200m. Flight is swift. Both sexes fond of over-ripe fruits and tree sap.

Male underside

Male upperside

Male upperside

Female upperside

Common Baron ■ *Euthalia aconthea* 5.5–8cm

DESCRIPTION Ground colour varies from plain brown to greenish. Pale spots on forewing are variable. Females of Sri Lankan subspecies *vasanta* have prominent white band across forewing. Subspecies *anagama* is from northern Western Ghats and peninsular India, while *garuda* is from North-east India. Other Indian subspecies are similar. **DISTRIBUTION** Throughout India and Sri Lanka to China and Sulawesi in mango groves. **HABITATS AND HABITS** Males are territorial and females are usually found in immediate vicinity of their host plant, the mango tree. Flight powerful, generally consisting of a flap and glide with wings held below the horizontal while gliding. Both sexes fond of over-ripe fruits, but also visit flowers.

Male upperside garuda

Male upperside anagama

Female anagama

Female vasanta

Male underside

Grand Duchess ■ *Euthalia patala* 8.5–12cm

Male underside

DESCRIPTION Sexes similar, except for pale spots across forewing, which are yellowish in males and white in females. **Green Duke** *E. sahadeva* (8–10.5cm; Nepal to Myanmar and west China) has differently shaped pale spots on forewing; female has white spots. **DISTRIBUTION** Pakistan to eastern Himalaya and Myanmar. **HABITATS AND HABITS** An inhabitant of dense forests of Himalayan Oak, at 1,000–2,800m. Single brood in May and June, when the butterfly is abundant, gliding like a ghost about shady ravines a metre or less above the ground. Settles frequently at wet sand, and both sexes are fond of over-ripe fruits and tree sap.

Male upperside

Male Green Duke

Female Green Duke

Redspot Duke ■ *Euthalia evelina* 7.5–10cm

DESCRIPTION Both sexes have prominent red spot on forewing. Northern subspecies *derma* is brown, South Indian subspecies *laudabilis* is green, and Sri Lankan subspecies *evelina* is olive-green. Females of southern Indian subspecies have larger pale area on upperside forewing. **DISTRIBUTION** Sri Lanka; peninsular India from Gujarat to Kerala along seaward face of Western Ghats and from Sikkim eastwards to Sulawesi. **HABITATS AND HABITS** An inhabitant of dense evergreen forest at low elevation. Never found away from forest. Flies low among the undergrowth in the deepest shade, becoming barely visible. Flight swift and powerful, consisting of a few flaps followed by a long glide. Both sexes attracted to over-ripe fruits and sap.

Male derma

Female derma

Male laudabilis

Female laudabilis

French Duke ■ *Euthalia franciae* 7.5–9cm

DESCRIPTION Sexes similar. A singular butterfly in India. **DISTRIBUTION** Central Nepal to northern Myanmar and southern China. **HABITATS AND HABITS** A forest insect, never found away from dense forest at moderate elevation. Flight powerful, a few flaps followed by a glide. Males fond of wet sand. Both sexes attracted to over-ripe fruits, dung and tree sap.

Male upperside

Underside

Gaudy Baron ■ *Euthalia lubentina* 6–8cm

Female underside

DESCRIPTION Red spots on wings are singular on Indian peninsula and Sri Lanka; in Assam very similar **Fruhstorfer's Baron** *E. malaccana* (Assam to Borneo) has been recorded, which differs from Gaudy in having largest white spot on forewing elongate in females. **DISTRIBUTION** Sri Lanka; peninsular India; Himalaya from Himachal Pradesh eastwards to the Philippines and Borneo. **HABITATS AND HABITS** Flight powerful, usually high around treetops in forests. Both sexes fond of over-ripe fruits and tree sap, though spring brood has also been recorded at flowers, perhaps due to lack of fruits in that season.

Female and Male Fruhstorfer's

Female upperside

Male upperside

Baronet ■ *Symphaedra nais* 6–7cm

Underside

DESCRIPTION Sexes similar. Species is singular on Indian subcontinent. White band on underside hindwing may or may not be present. **DISTRIBUTION** Endemic to Indian subcontinent and Sri Lanka. **HABITATS AND HABITS** Common in low-elevation deciduous and *Shorea robusta* forests. Flight powerful, and near the ground. Settles frequently to bask on the ground. Both sexes attracted to over-ripe fruits, dung and tree sap.

Upperside

Underside

Northern Junglequeen ■ *Stichohpthalma camadeva* 12.5–15cm

DESCRIPTION Sexes similar. Upperside is singular. Underside is also distinct. **Chocolate Junglequeen** *S. nourmahal* (9.5–10.5cm; Sikkim eastwards to Nagaland) is similar, but easily distinguished because it is orange on upperside and rather different on underside. **DISTRIBUTION** Eastern Nepal to North-east India, Myanmar to Thailand. **HABITATS AND HABITS** This genus is generally found on hills in dense jungle at moderate elevation. On the wing during rainy season. Active at dawn and dusk, when both sexes visit over-ripe fruits and tree sap.

Upperside

Underside

Underside Chocolate

Common Faun ■ *Faunis canens* 6.5–7.5cm

DESCRIPTION Sexes similar; on underside hindwing, dark middle line is joined to basal line to form a rough 'U'. In similar **Large Faun** *F. eumeus* (8–9.5cm; Arunachal Pradesh to Myanmar and China), middle line is joined to line along margin, again forming a 'U'. Both species more or less the same on upperside. **DISTRIBUTION** Sikkim to Borneo. **HABITATS AND HABITS** Inhabits dense broadleaved forests at low elevation. Active in deepest shade of the forest, or at dawn and dusk. Flight powerful, and butterflies settle frequently. Both sexes attracted to over-ripe fruits and tree sap.

Underside

Upperside

Underside Large

Southern Duffer ■ *Discophora lepida* 8–9cm

DESCRIPTION Sexes similar, but female has larger blue spots on upperside than male; undersides as illustrated. Similar **Common Duffer** *D. sondaica* (8–9cm; central Nepal to Myanmar and the Philippines) is variable on upperside, with or without blue or yellow spots; it never has a broad yellow band; female has angular hindwing. **DISTRIBUTION** Endemic to Sri Lanka and from Goa to Kerala along Western Ghats. **HABITATS AND HABITS** Active at dusk, when males venture into clearings or forest edges, where they fly about at around 4m above the ground in the gloaming. Flight fast and powerful, with males settling frequently while patrolling their beats. Females presumably pass by and are mated, but are generally found within dense forests. Both sexes attracted to over-ripe fruits and tree sap.

Male underside

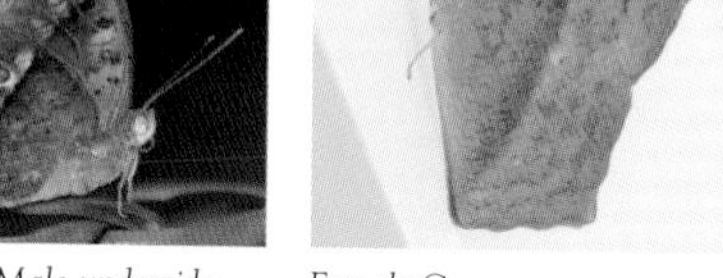

Female Common

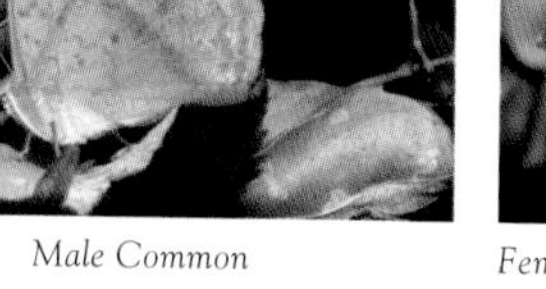

Male Common

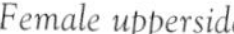

Female upperside

Common Palmking ■ *Amathusia phidippus* 10–12.5cm

DESCRIPTION Sexes as illustrated. Large butterfly singular in South India. Very similar **Andaman Palmking** *A. andamanensis* occurs on Andaman Islands. **DISTRIBUTION** Kerala; Myanmar to Malaysia, the Philippines and Sulawesi. **HABITATS AND HABITS** Active at dusk in coconut groves. Occasionally attracted to artificial light. Flight powerful and it keeps to shady thickets, so it is not easy to see. Both sexes fond of over-ripe fruits and tree sap.

Male upperside

Female upperside

Underside

Jungleglory ■ *Thaumantes diores* 9.5–11.5cm

DESCRIPTION Sexes similar. Upperside reminiscent of Great and Danaid Eggfly (see p. 118) males, but underside immediately distinguishes this species from them. **DISTRIBUTION** Eastern Nepal to Myanmar, Thailand and southern Vietnam. **HABITATS AND HABITS** Inhabits dense forest at low elevation. On the wing at dawn and dusk, and does not take flight during the day unless disturbed. Both sexes fond of over-ripe fruits and tree sap.

Upperside

Underside

Common Palmfly ■ *Elymnias hypermnestra* 6–8cm

DESCRIPTION Male's upperside and underside of both sexes distinctive. Female's upperside and flight closely resemble those of Plain and Common Tigers (see p. 121), but she has a unique, sharply angled forewing-tip. South Indian subspecies *caudata* has tails to hindwing (southern Maharashtra to Kerala). **DISTRIBUTION** Punjab to North-east India, peninsular India and Sri Lanka to Vietnam and Indonesia. **HABITATS AND HABITS** Common in palm groves. Active in shady thickets and forests at dusk. Keeps to shade during daytime. Males territorial and occasionally visit wet sand. Both sexes fond of over-ripe fruits. Females perfectly mimic flight and behaviour of Plain and Common Tigers, and are difficult to distinguish on the wing even by experienced eyes.

Female undularis

Female undularis

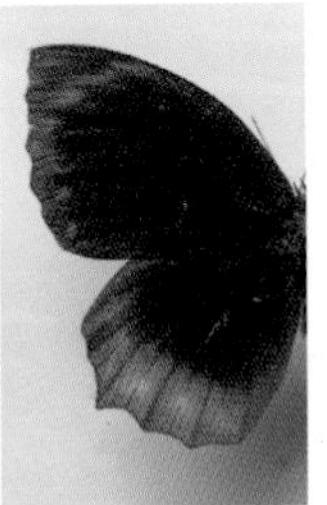

Male undularis

Male caudata

Dusky Diadem ■ *Ethope himachala* 6–8.5cm

DESCRIPTION Sexes similar. Female has rounder wings than male. Row of eyespots along edges of wings on both surfaces is singular. **DISTRIBUTION** Sikkim to North-east India and Thailand. **HABITATS AND HABITS** An inhabitant of dense forest in regions of heavy rainfall at low elevation. Flight weak and ragged, and butterfly settles frequently on leaves with wings outspread. Not attracted to wet sand. Though it generally keeps within the forest, it ventures on to paths and along streams.

Underside

Male upperside

Female upperside

Common Evening Brown ■ *Melanitis leda* 6–8cm

DESCRIPTION Sexes similar. Wet-season form's underside singular, mottled with row of eyespots. Dry-season form's underside colour and pattern highly variable. Shape of forewing distinctive. **Dark Evening Brown** *M. phedima* (6–8.5cm; peninsular India; Jammu and Kashmir to Japan and the Philippines), which is equally variable in dry-season form, has less pronounced forewing apex and no orange on upperside forewing. **DISTRIBUTION** Throughout India and Sri Lanka; Africa through Asia to Australia. **HABITATS AND HABITS** Active in evenings. Flight jerky and the insect settles frequently. When threatened, it settles among dry leaves and leans over to lie flat. Both sexes fond of over-ripe fruits and tree sap.

Upperside

Dark

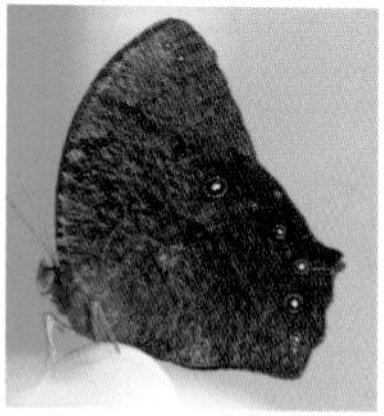

Dry- and wet-season Dark Evening

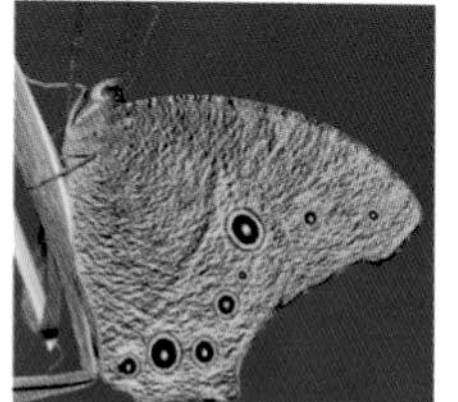

Wet- and dry-season underside

Travancore Evening Brown ■ *Parantirrhoea marshalli* 5.5–6.5cm

DESCRIPTION Sexes as illustrated. A singular species. This unique genus and species is endemic to India and was little known until it was bred by C. Susanth and its habits were understood. **DISTRIBUTION** Karnataka to Kerala along seaward face of Western Ghats. **HABITATS AND HABITS** Only found in cane brakes in dense forest at low elevation at a few locations, but it cannot be said to be endangered because most of the known localities are within well-protected areas. Common in suitable localities. Active at dawn and dusk, spending the remaining time skulking in cane thickets. At dusk, flight is rapid, with males patrolling a beat for females. Both sexes fond of over-ripe fruit and tree sap, especially toddy.

Male upperside

Female upperside

Underside

Banded Treebrown ■ *Lethe confusa* 5–6.5cm

DESCRIPTION Sexes similar. Underside hindwings with eyespots not disintegrated. Upperside has white band across forewing, which is angled downwards at lower end. Similar **Straight-banded Treebrown** *L. verma* (5–6cm; Jammu and Kashmir to Taiwan and Thailand) has straight and even white band across forewing. Wings rounder than in Banded, and hindwing tail not as prominent. **DISTRIBUTION** Pakistan to North-east India, Hainan and Indonesia. **HABITATS AND HABITS** A forest butterfly, never found in open country. Ascends Himalaya to 1,800m. Flight swift and jerky, settling frequently. Both sexes fond of over-ripe fruits and tree sap. The species has been recorded recently on flowers.

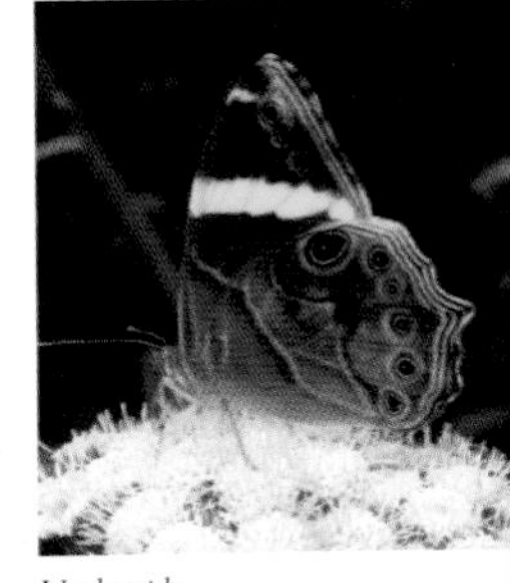
Underside

Upperside

Straight-banded

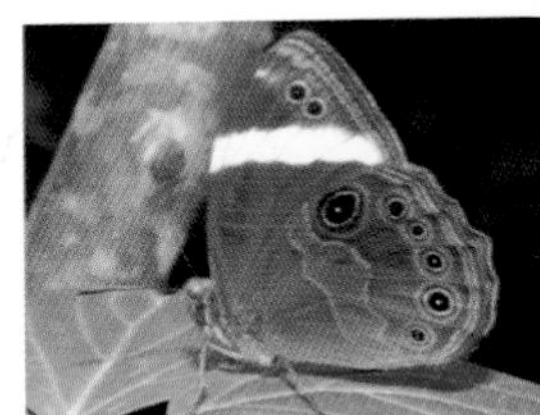
Straight-banded

Bamboo Treebrown ■ *Lethe europa* 6.5–8cm

DESCRIPTION Female has prominent white band across forewing, which male lacks. In both sexes, on underside, narrow white band across middle of both wings is straight. In similar **Tamil Treebrown** *L. drypetis* (6.5–7cm; endemic to Sri Lanka and from Odisha and Goa to Kerala) white line is irregular. **DISTRIBUTION** Peninsular India south of Gujarat and Odisha; Punjab to Japan and the Philippines. **HABITATS AND HABITS** Ascends to 2,000m in Himalaya, but most common at low elevation. Found in dense, shady forest with bamboo undergrowth. Flight swift and jerky. Most active at dawn and dusk. Neither sex visits flowers, but both sexes are fond of fruits, dung and tree sap.

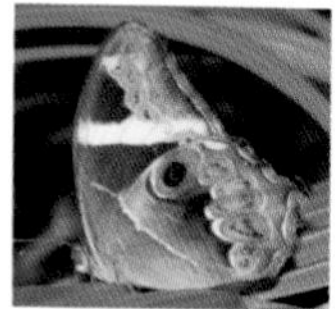
Female underside

Male underside

Female upperside

Male underside Tamil

Common Treebrown ■ *Lethe rohria* 5.8–6.5cm

DESCRIPTION On underside hindwing, three eyespots are disintegrated (red arrow). Females of Himalayan subspecies *rohria* have white band across forewing, which in southern Indian subspecies *neelgheriensis* is divided into white spots. **DISTRIBUTION** Sri Lanka; peninsular India as far north as Rajasthan and West Bengal; Afghanistan to Pakistan, eastwards to North-east India, Taiwan and Indonesia. **HABITATS AND HABITS** Though a forest insect, both sexes venture into gardens and plantations in search of food sources. Often appears in numbers after rainy season. Flight rapid and jerky, with butterflies settling frequently. Territorial behaviour by males has not been noted. Both sexes fond of over-ripe fruits and occasionally visit flowers.

Female rohria

Male rohria

Female neelgheriensis

Common Wall ■ *Lasiommata schakra* 5.5–6cm

DESCRIPTION Sexes similar, but male has reduced orange markings on upperside forewing. Singular in outer Himalayan ranges, but there are two similar species in main Himalayan range. **DISTRIBUTION** Endemic to western Himalaya, from Pakistan to western Nepal. **HABITATS AND HABITS** Found only around cliffs and rock faces, and on steep grassy hillsides at 1,200–3,000m. Flight weak and jerky, with butterflies settling frequently. Both sexes attracted to flowers.

Underside

Female upperside

Male upperside

Tiger Brown ■ *Orinoma damaris* 7.5-8cm

DESCRIPTION Sexes similar. The yellow spot at the base of the forewing cell is unique to this butterfly. In flight it mimics Great Blackvein A. *agathon* (see p. 42) but there is no yellow mark on the upperside forewing of the Great. **DISTRIBUTION** Areas of heavy rainfall along the Himalaya from Jammu and Kashmir to North-east India and Myanmar. **HABITATS AND HABITS** Found along the Himalaya at moderate elevation, generally at 1,200–1,600m, in dense broadleaved forests. Never found away from shady forest, and prefers ravines. Males patrol a beat for days at a time. Both sexes fond of tree sap and over-ripe fruits. Summer brood mimics Great Blackvein, which is much more numerous than Tiger Brown.

Upperside

Underside

Gladeye Bushbrown ■ *Mycalesis patnia* 4–4.5cm

DESCRIPTION Sexes similar. Large eyespot on forewing is singular. Sri Lankan subspecies *patnia* brighter brown than Indian subspecies *junonia*. **DISTRIBUTION** Endemic to Sri Lanka and hills of southern India as far north as Maharashtra. **HABITATS AND HABITS** Inhabits forested hills where it flutters about in the undergrowth. Not found away from forests and well-wooded areas. Usually occurs singly. Fond of over-ripe fruits and tree sap.

Upperside patnia

Underside patnia

Underside junonia

Tamil Catseye ■ *Zipaetis saitis* 6–6.5cm

DESCRIPTION Sexes similar. Arrangement of eyespots on underside and white bar across each wing on upperside are singular. **Dark Catseye** *Z. scylax* (5.5–6.2cm; Sikkim to North-east India and southern China) is similar, but lacks white bars on upperside. **DISTRIBUTION** Endemic to southern India along seaward face of Western Ghats from Karnataka to Kerala. **HABITATS AND HABITS** An inhabitant of dense evergreen forests at low elevation. Keeps to shade during daytime, becoming active at dusk. Flight weak and fluttering, with butterflies settling frequently within bushes. Probably attracted to over-ripe fruits.

Upperside

Underside Dark

Underside

Ringed Argus ■ *Callerebia annada* 6–7cm

DESCRIPTION Sexes similar. On underside hindwing, there are two diffuse dark bands, one across centre of wing, the other leading into two eyespots at bottom of wing. In similar **Hybrid Argus** *C. hybrida* (5–6cm; Himachal Pradesh to central Nepal) there are narrow dark lines instead of diffuse bands, and outer band does not lead into eyespots at bottom of wing. In **Moore's Argus** *C. orixa* (5.5–6cm; Meghalaya and Nagaland) orange ring around upperside forewing eyespot is broader than in either of the two preceding species. **DISTRIBUTION** Pakistan to Arunachal Pradesh. **HABITS** Flight consists of a series of hops. Both sexes fond of flowers and wet mud in oak forests above 1,000m.

Underside Hybrid

Upperside

Underside

Common Argus ■ *Callerebia nirmala* 4.5–5.5cm

DESCRIPTION Sexes similar. No bands or lines on underside hindwing; at most, a faintly marked dark band across middle of wing. Number and size of eyespots on both surfaces of hindwing variable. Similar **Pallid Argus** *C. scanda* (5–7.6cm; Jammu and Kashmir to Bhutan) suffused with white on underside hindwing, and has pale borders to wings on upperside. **DISTRIBUTION** Afghanistan to western Nepal. **HABITATS AND HABITS** Generally found in forest glades and along paths at 1,000–2,800m in Himalaya. **HABITATS AND HABITS** Flight a series of erratic hops. Males congregate at wet mud, and both sexes fond of flowers.

Upperside Pallid

Upperside

Underside

Underside Pallid

Common Threering ■ *Ypthima asterope* 3–3.7cm

DESCRIPTION Sexes similar. In wet-season form, it is unmistakable, with uppermost eyespot on underside hindwing being very small and no dark bands across hindwing. In dry-season form (not illustrated), there may or may not be up to two obscure dark bands across underside hindwing. **Lesser Threering** Y. *inica* (3–3.4cm; Pakistan to Assam) lacks all dark bands across underside wings in wet-season form. Uppermost eyespot on hindwing much larger than in Common. Dry-season form may or may not have dark bands on underside forewing between eyespot and wing margin, as well as up to two dark bands across middle of hindwing. **DISTRIBUTION** Assam westwards to Baluchistan, Lebanon and Africa. **HABITATS AND HABITS** Flight weak, hopping and near the ground. Inhabits open areas in drier areas. Both sexes fond of flowers.

Upperside Lesser

Underside wet season

Underside Lesser

Common Fourring ■ *Ypthima hubneri* 3–4cm

Wet season

DESCRIPTION On underside hindwing, there are four eyespots. Wet-season form has well-developed eyespots on underside hindwing, while they are reduced to dots in dry-season form. **White Fourring** Y. *ceylonica* (3–3.5cm; Sri Lanka to West Bengal) has variable number of eyespots on underside hindwing and half upperside hindwing white. **Kashmir Fourring** Y. *kasmira* (3.5–4cm; northern Pakistan to central Nepal) is identical to Common, but lacks all dark bands on underside. **DISTRIBUTION** Throughout India to China and Thailand. **HABITATS AND HABITS** A forest insect, found hopping along grassy verges of roads and in open country, always near the ground. Both sexes fond of low-growing flowers.

White

Female upperside

Dry season

Wet season Kashmir

Common Fivering ■ *Ypthima baldus* 3.2–4.8cm

DESCRIPTION Upperside forewing has yellow ring on paler area than rest of wing. Pale area defined by dark band. On upperside hindwing, always a dark line across middle of wing. Five eyespots on underside hindwing (the lowest two counting as one); centres of lower four spots not in line. Similar **Baby Fivering** *Y. philomela* (2.5–3.5cm; Kerala to Maharashtra, West Bengal and Myanmar) is smaller, lacks line across upperside hindwing and all dark bands on underside. Uppermost eyespot on underside hindwing usually very small. **Nilgiri Jewel Fourring** *Y. striata* (3.5–4.5cm; Kerala and Tamil Nadu) distinguished by prominent dark bands across underside. **DISTRIBUTION** Throughout India to Indonesia, Japan and Ussuri. **HABITATS AND HABITS** An ubiquitous insect, it ascends hills to more than 2,000m and is found in forests, fields and gardens. Hopping flight takes it safely through bushes. Both sexes fond of flowers.

Upperside

Nilgiri Jewel

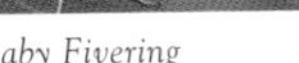

Baby Fivering

Dry season

Wet season

Moore's Fivering ■ *Ypthima nikaea* 4.5–5cm

DESCRIPTION Sexes similar. On underside hindwing, distinguished from **Himalayan Fivering** *Y. sakra* (4.8–5.5cm; Himachal Pradesh to North-east India to China) by uppermost two eyespots on hindwing being separated by yellow band (red arrow), while in Himalayan they are joined together. **DISTRIBUTION** Jammu and Kashmir to Nepal. **HABITATS AND HABITS** A common species in oak forest above 1,400m. Both sexes prefer to fly about in shady ravines, but do venture out to cross open spaces. Flight more powerful than that of smaller members of the genus. Males gather in numbers at wet sand. Both sexes fond of flowers.

Underside

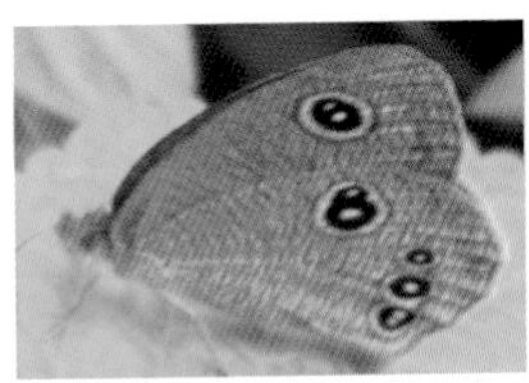

Underside Himalayan

Upperside

▪ Photo credits ▪

(Photos are denoted by a page number , a row number from the top in square brackets and L = left, C= centre, ML = Middle Left, MR = Middle Right and R = right.)

16 [2L, 3R & 4R] Resmi Varma; [2R & 3L] Milind Bhakare; 17 [1R, 2L & 4L] C. Susanth; [2R] Tharaka Priyadarshana; [3R] Himesh Dilruwan Jayasinghe; [4C] Prashanth S.N.; 18 [3L] Ganesh Hegde; [4R, 5R] Ngangom Aomoa; [5L, 5C] Atanu Bora; 19 [2L]Deep Brahma; [2R, 4C] Adam Cotton; [3R] Atanu Bora; [4R] Sunny Chir; 20 [1L, 1R] Milind Bhakare; [1C] Atanu Bose; [2L] Atanu Bora; [3L] Resmi Varma; [4R] Sarab Seth; 21 [1L] Milind Bhakare; [1R] Sarab Seth; [2R, 3R, 4L & 4R] Milind Bhakare; 22 [1L] Resmi Varma; [2R] Anjali Prasad; [3L, 3C & 3R] Ngangom Aomoa; [2ML] Jis Sebastian; [2MR] Sunny Chir; [2R & 4L] Rachit Singh; [3R] Ganesh Mani Pradhan; 24 [2L] Ngangom Aomoa; [2R] Atanu Bose; [3L]Ashok Sengupta; [4L] C. Susanth; [4C & 4R] Mymoon Mogul; 25 [1L] Milind Bhakare; [1R] C. Susanth; [2R] Atanu Bose; 26 [2L] Sandex Verghese; [2R] C. Susanth; 27 [1L] Chinmayi S.K.; [1R, 3L & 3R] Milind Bhakare; 28 [1L] Vonchano Nguille; [2L] Harmenn Huidrom; [2C] Zdenek Hanc; [2R] Nosang Muringa Limboo; [3L & 3R] Milind Bhakare; [3C]Deep Brahma; 29 [1L] C. Susanth; [1R] Milind Bhakare; [2L & 2C] Alok Mahendroo; [2R] Antonio Giudici; 30 [1L] Pius Smetacek; [1ML] Subhajit Mazumder; [1MR, 1R] Nikhil Bhopale; [3L] C. Susanth; [3R] Atanu Bose; 31 [1L, 1R, 3L & 3R] Milind Bhakare; [2L] Sarab Seth; [2R] Adam Cotton; 32 [1L] Milind Bhakare; [1M] Adam Cotton; [1R] Resmi Varma; [2L] Atanu Bose; [2M & 2R] Sunny Chir; 33 [1R & 2L] Ngangom Aomoa; [2R & 3R] Tatsuki Watanabe; [4L] Antonio Giudici; [4R] Harshavardhan Huidrom; 34 [1L] Tshetsholo Naro; [1R] Tatsuki Watanabe; [3L] Shashank Pathour; [3R] Basil Wirth; 35 [1R] Emmanuel Theophilus; [3L & 3C] Heiner Zeigler; [3R] Toshihiko Katayama; 36 [3R] Ngangom Aomoa; [4R & 5R] Sujeeva Gunasena; [5L] Chinmayi S.K.; 38 [1C] Ngangom Aomoa; [2C] Milind Bhakare; [2R] Kasun L. Wickramasingha; 39 [1L] Samhita Kashyap; [1C] Milind Bhakare; 40 [4MR] Milind Bhakare; [[4R] Ngangom Aomoa; 42 [2C] Tashi R. Ghale; [2R] Ngangom Aomoa; 43 [1L] Milind Bhakare; [1R] Sonam Dorji; [2L & 2C] Toshihiko Katayama; [2R] Milind Bhakare; 44 [1L] Sarab Seth; [2R] Ngangom Aomoa; [3L] Pius Smetacek; [4ML] Muhammad Ackram; 45 [1L] Ngangom Aomoa; [1C] Rakesh Khedwal; [1R] Ramya HR; [3L] Ngangom Aomoa; 46 [1L] Divakar Thombre; [2C] Milind Bhakare; [2R] C. Susanth; [3L] Divakar Thombre; [4L] Milind Bhakare; 47 [1L & 1R] Milind Bhakare; [1C] Sarab Seth; [2R, 3C & 3R] Milind Bhakare; [3L] Divakar Thombre; 48 [1L & 1R] Milind Bhakare; [1C] Chinmayi S.K.; [2L] Milind Bhakare; [3L] Ngangom Aomoa; [3C] Chinmayi S.K.; [3R] Kishen Das; 49 [1R & 2R] Divakar Thombre; [2L, 2C, 4C & 4R] Ngangom Aomoa; [3R & 4L] Milind Bhakare; 50 [1L, 2ML] C. Susanth; [2L] Divakar Thombre; [2MR] Samhita Kashyap; [2R] Chinmayi S.K.; [3L] C. Susanth; [4L] Atanu Bose; [4ML] Ngangom Aomoa; [4R] Sudarshana Borah; 51 [1R] Chinmayi S.K.; [2L] Tanvir Ahmed Shaikot; [2ML; 3R & 4R] Milind Bhakare; [2MR] Subhajit Mazumder; [2R] Haneesh Km; [4L] Vijay Anand Ismavel; [4ML] Ajith Unnikrishnan; [4MR] Ngangom Aomoa; 52 [2L, 3R & 4L] Milind Bhakare; [2M] Atanu Bose; [2R] Bednedhi Dhakal; [3L; 4R] Ishara H. Wijewardhane; 53 [1L] Ngangom Aomoa; [1C] Milind Bhakare; [1R] Mayuresh Kulkarni; [3L & 3R] Milind Bhakare; 54 [1L] Milind Bhakare; [2L, 4L] Ngangom Aomoa; [2R] Sonam Dorji; [4C; 4R] Atanu Bose; 55 [1R, 2R] Tom & Victor Schneider; [2C] Sonam Dorji; [3R, 4R] John Christis; [4L] Ngangom Aomoa; [4C] Lallawmsanga Tetea Zeon; 56 [1L] Rajashree Bhuyan; [1R] Sonam Dorji; [3C] Milind Bhakare; [3R] Atanu Bose; 57 [1R, 2R] Milind Bhakare; [2L] Haneesh Km; [3R, 4R] Ngangom Aomoa; [4L] Sunny Chir; 58 [2L] Milind Bhakare; [2R] Divakar Thombre; [3L] Tharaka S. Priyadarshana; [4L] Chinmayi S.K.; [4R] C. Susanth; 59 [1R] Divakar Thombre; [2L] Mohit Patel; [2C] Pradip Patade; [2R] Nelson Rodrigues; [3R, 4L] Milind Bhakare; [4C] Sarab Seth; [4R] Avinash Sant; 60 [2L] Milind Bhakare; [4L & 4R] Subhajit Mazumder; 61 [2L] Atanu Bose; {2ML} Subhajit Mazumder; [3R, 4L] Tharaka S. Priyadarshana; 62 [2L] Luvjoy Choker; [2C] Chungchung Chhetri; [4L] Milind Bhakare; [4C, 4R] Md. Samsur Rahman; 63 [1R, 4L, 4ML] Milind Bhakare; [2L] C. Susanth; [2R] Divakar Thombre; [4R] Anjali Prasad; 64 [1L, 2L, 2R] Parixit Kafley; [3L] Milind Bhakare; [3C] Sunny Chir; 65 [2R] Krinal Jani; 66 [2R,3R] Ashwini Bhatia; [3C] Shankar Kumar; 67 [1R, 2R] Kishen Das; [2L, 3C, 3R] Milind Bhakare; 68 [1L] Pratiksha Patel; [1C, 1R] Milind Bhakare; [3R] Rajashree Bhuyan; 69 [1L, 1ML] Motoki Saito; [1MR, 1R] Sonam Dorji; [2L] Divakar Thombre; [2C] Krupa George; 70 [1R, 2R] Rachit Singh; [2L] Rajashree Bhuyan; [3R] Milind Bhakare; 71 [1L] Rajashree Bhuyan; [1C] Milind Bhakare; [2R] Sonam Dorji; [3L] Ganesh Hegde; [3C] Chinmayi S.K.; [3R] Shankar Kumar; 72 [1L] Milind Bhakare; [1C] C. Susanth; [1R, 2L, 3R] Divakar Thombre; [3L] Atanu Bose ; 73 [1R] Atanu Bora; [2L] Chinmayi S.K.; [2C] C. Susanth; [2R, 3C]Milind Bhakare; [3L] Rajashree Bhuyan; [3R} Shankar Kumar; 74 [1C, 1R] Kishen Das; [2L] Divakar Thombre; [3L] C. Susanth; [3ML, 3MR, 3R] Milind Bhakare; 75 [1L, 1C, 2L, 2R] Milind Bhakare; 76 [1L; 2L, 2R, 4L] Milind Bhakare; [2C] Atanu Bose; [3L] Leslie Day; [4C] Divakar Thombre; 77 [1R] Sonam Dorji; [2R] Amlan Mitra; [3L] Divakar Thombre; [3R] Yuwaraj Gurjar; 78 [1L, 1R] Milind Bhakare; [2L] Atanu Bose; [2R] Blaise Pereira; [3L] Chandrakant Ashtekar; [3C] Pratiksha Patel; [3R] Krupa George; 79 [1R, 2L, 2C, 2R, 3R] Milind Bhakare; [3L, 4L, 4R] Kishen Das; 80 [1L] Milind Bhakare; [2C] Kishen Das; [2R, 3R] Divakar Thombre; 81 [2L] Krupa George; [3L, 4R] Ishara H. Wijewardhane; [3R] Milind Bhakare; [4L] Krinal Jani; [4C] Kishen Das; 82 [1L, 1R] Kishen Das; [2L, 2R]

Samhita Kashyap; [2C] Milind Bhakare; [3L] C. Susanth; [3C] Chinmayi S.K.; [3R] Divakar Thombre; 83 [1R] Sonam Dorji; [2ML, 2R] Samhita Kashyap; [3R] Kishen Das; [4L, 4ML, 4MR] Milind Bhakare; 84 [1L] Sarab Seth; [1C] Milind Bhakare; [1R] Chinmayi S.K.; [2C] Kishen Das; [2R] Divakar Thombre; 85 [1L] Milind Bhakare; {1R] C. Susanth; [2R] Luvjoy Choker; [3L] Kishen Das; [3R, 4C] Milind Bhakare; [4L] Samhita Kashyap; [4R] Yuwaraj Gurjar; 86 [1L] Chinmayi S.K.; [1ML] Atanu Bose; [1C; 1MR] Milind Bhakare; [1R] Kishen Das; [3L, 4L] Samhita Kashyap; [3R] Pratiksha Patel; 87 [1L] Divakar Thombre; [2R] Milind Bhakare; [3C, 3R] C. Susanth; 88 [2L] Matrika Sharma; [2C, 3L] Milind Bhakare; [2R] Ngangom Aomoa; 89 [1R] Milind Bhakare; [2L, 2R] Atanu Bose; 90 [1L, 3L, 3C, 3R] Milind Bhakare; [1C] Subhajit Mazumder; [1R] Tanvir Ahmed Shaikot; 91 [1L] Ngangom Aomoa; [1R] C. Susanth; [2R, 3L] Milind Bhakare; 92 [1L, 1R, 2L, 2R] Milind Bhakare; 93 [2L] Milind Bhakare; [2C] Divakar Thombre; 94 [2L] Milind Bhakare; [2R] Divakar Thombre; [3L] C. Susanth; [3R] Milind Bhakare; 95 [1L, 1C] Milind Bhakare; [2R] C. Susanth; 96 [1C] Rajashree Bhuyan; [2L, 2R] Milind Bhakare; 97 [1R, 2C, 4L, 4R] Ngangom Aomoa; [2L, 2R] Ganesh Mani Pradhan; [3R] Deep Brahma; 98 [1L, 1R] Nuwan Chaturanga; [2L, 2C] Milind Bhakare; [2R] C. Susanth; [3L] Sunil Bhoite; 99 [2R] C. Susanth; [3R, 4R] Milind Bhakare; [4L, 4C] Ngangom Aomoa; 100 [1L, 1C] Jatishwor Singh Irungbam; [1R] Sunny Chir; [2L, 3L] Ngangom Aomoa; [3R] Matrika Sharma; 101 [1L, 1R] Ashwini Bhatia; [2R, 3L] Divakar Thombre; [3C] Alka Vaidya; [3R] Madan Shrestha; [2L] Ngangom Aomoa; [3L] Matrika Sharma; 103 [1L] Matrika Sharma; [1C, 1R] C. Susanth; [2R] Divakar Thombre; [3R] Munir Ahmed Khan; [4L] Ngangom Aomoa; [4R] Atanu Bora; 104 [2L] Atanu Bose; [3L] Sonam Dorji; 105 [1L] Ngangom Aomoa; [1R] Sonam Dorji; [2R, 4L] Prashanth S.N.; [3R] Ngangom Aomoa; [4C] Milind Bhakare; 106 [1L] Saurabh Nandi; [2L, 2R, 3L, 3C] Ngangom Aomoa; [3R] Alka Vaidya; 107 [1R, 2R] Atanu Bose; [2L] Ravi Bhambure; [2C] C. Susanth; 108 [3R] Ishara H. Wijewardana; [4L] Ngangom Aomoa; [4C] Nilanjan Bhattacharya; [4R] C. Susanth; 108 [1L, 2R, 3L, 3R] Subhajit Mazumder; [1R, 2ML] Saurabh Nandi; 109 [1L, 1R, 2L, 2R] Milind Bhakare; [2C] Sonam Dorji; [4L, 4C] Krupa George; 110 [1L] Sarab Seth; [1R] Shankar Kumar; [2L, 3L] Ngangom Aomoa; [3R] Luvjoy Choker; 111 [1L, 1R, 2L, 2R] Ngangom Aomoa; [3R] Sarab Seth; 112 [1L, 2R] Rachit Singh; [2L, 2C] Milind Bhakare; [3L, 3R] Ngangom Aomoa; 113 [2R] Muhammad Ackram; [3R, 4L] Resmi Varma; [4C] Sachindra Umesh; [4R] Ngangom Aomoa; 114 [1R] Soibam Baleshwor; [3R] Christopher A. Rickards; 115 [1L, 2L, 2R] Milind Bhakare; [1C] Samhita Kashyap; [1R] Mohit Patel; 116 [1L] Sunil Bhoite; [2L, 2R, 3L, 3C] Milind Bhakare; [3R] Mohit Patel; 117 [1R, 2C] Milind Bhakare; [2R] Resmi Varma; 118 [1L, 2L, 2C, 3L, 4L, 4C, 4R] Milind Bhakare; [2R] C. Susanth; 119 [1R] Tharaka S. Priyadarshana; [2L] Kishen Das; [2R] Motoki Saito; [3L] Ngangom Aomoa; [3C] Jatishwor Singh Irungbam; 120 [1L] C. Susanth; [1C] Ngangom Aomoa; [1R] Kishen Das; [2L, 2R] Atanu Bora; 121 [2L] Sunny Chir; [2MR] Krinal Jani; [2R] Divakar Thombre; [4L, 4C] Nelson Rodrigues; 122 [1L] Resmi Varma; [2L] Tharaka S. Priyadarshana; [3L] Ishara H. Wijewardana; [3R] C. Susanth; 123 [1R, 2L] Milind Bhakare; [2C] Ngangom Aomoa; [3R] Chinmayi S.K.; [4R] Resmi Varma; 124 [1L] Milind Bhakare; [1C] C. Susanth; [1R] Tharaka S. Priyadarshana; [2R] Sachindra Umesh; 125 [1R] Atanu Bose; [2L] Subhajit Mazumder; [2C] Chinmayi S.K.; [2R] Samhita Kashyap; [3L] Milind Bhakare; [3C, 3R] Ngangom Aomoa; 126 [1L, 1C, 1R] Milind Bhakare; [3L] Ngangom Aomoa; [3R] Atanu Bose; 127 [2L] Subhajit Mazumder; [2R] Milind Bhakare; 128 [1L] Neeraj Sharma; [1C, 1R] Ngangom Aomoa; [3R] Sachindra Umesh; 129 [1L, 2L, 2R] Milind Bhakare; [2C] Atanu Bose; 130 [2L] Milind Bhakare; [3R] C. Susanth; 131 [1L, 2R, 3C] Ngangom Aomoa; [1R] Divakar Thombre; 132 [1L] Milind Bhakare; [1C] Divakar Thombre; [2L, 3L] Nuwan Chaturanga; [3R] Chinmayi S.K.; 133 [3R, 4L, 4R] Milind Bhakare; 134 [1L, 1C, 1R] Milind Bhakare; [3L, 3R] Sachindra Umesh; 135 [1L] Ngangom Aomoa; [1ML] Pradip Patade; [1MR, 1R] Milind Bhakare; [2ML, 2MR] Jatishwor Singh Irungbam; [2R] Divakar Thombre; 136 [1C, 3L, 3R] Ngangom Aomoa; 137 [2C] Rajashree Bhuyan; [3L, 3C] Atanu Bose; [3R] Sonam Dorji; 138 [1L, 1R] Divakar Thombre; [1C] Milind Bhakare; [2R, 3MR] Luvjoy Choker; [3L} Shankar Kumar; [3ML] Divakar Thombre; 139 [1L, 1C, 1R, 2C] Milind Bhakare; 140 [1L] Chinmayi S.K.; [2L] Resmi Varma; [2ML] C. Susanth; [2MR] Tania Khan; [3L] Milind Bhakare; [3R] Ramya H.R.; 141 [1L] Atanu Bose; [1R] Ngangom Aomoa; [2R] Lallawmsanga Tetea Zeon; [3L] Milind Bhakare; [3C, 3R] Atanu Bose; 142 [2L, 2ML, 2R] Milind Bhakare; [2MR] Sujeeva Gunasena; [3L] Ashwini Bhatia; [4C] Sonam Dorji; [4R] Matrika Sharma; 143 [1R, 2L, 3L, 3R] Ngangom Aomoa; [2C] Makarand Kulkarni; [2R] C. Susanth; 144 [1L, 2C, 2R] Milind Bhakare; [2L] Antonio Giudici; [3L] Divakar Thombre; [4L] C. Susanth; [4R] Kishen Das; 145 [1L, 1C] Kuschel Gurung; [2L] Ngangom Aomoa; [2C, 2R] Milind Bhakare; 146 [1L, 2R, 3L, 3C, 3R] C. Susanth; [1R] Luvjoy Choker; 147 [1L] Divakar Thombre; [1R] Ngangom Aomoa; [2L] Divakar Thombre; [2R] Milind Bhakare; 148 [1L] Divakar Thombre; [1R] Ngangom Aomoa; [2L] Atanu Bose; [3L, 3MR, 3R] Milind Bhakare; [3ML] Lallawmsanga Tetea Zeon; 149 [1L, 1C, 1R] C. Susanth; [3L] Atanu Bose; [3C] Jatishwor Singh Irungbam; 150 [1L] Milind Bhakare; [2C] Chinmayi S.K.; [2R] Divakar Thombre; [3R] C. Susanth; 151 [1L] Atanu Bose; [2L] Ngangom Aomoa; 152 [1L, 1C] Sachindra Umesh; [1R] Resmi Varma; [2L, 3R] C. Susanth; [3L] Ngangom Aomoa; 154 [1l, 1C, 1R] Milind Bhakare; [2L, 3ML] Pramita Roy; [3L] Arya Meher; [3R] Muhammad Ackram; 155 [2L] Pavendhan Appavu; [2ML, 2MR, 2R] Milind Bhakare.

Excerpted from: Varshney, R. K. & Smetacek, P. (eds) 2015. *A Synoptic Catalogue of the Butterflies of India*. Butterfly Research Centre, Bhimtal and Indinov Publishing, New Delhi. ii+261 pp., 8 pl.

Superfamily: Papilioniodea
Family: Papilionidae, Hesperiidae, Pieridae, Riodinidae, Lycaenidae, Nymphalidae
Subfamily: Papilioninae, Parnassiinae, Coeliadinae, Eudaminae ,Pyrginae, Heteropterinae, Hesperiinae, Coliadinae, Pierinae, Riodininae, Curetinae, Poritinae, Miletinae, Lycaeninae, Aphnaeinae, Theclinae, Polyommatinae, Danainae, Calinaginae, Charaxinae, Amathusinae, Satyrinae, Limenitidinae, Heliconiinae, Biblidinae, Apaturinae, Cyrestinae, Nymphalinae, Acraeinae, Libytheinae,
Tribe: Troidini, Leptocircini, Teinopalpini, Zerynthiini, Tagiadini, Celaenorrhini, Carcharodini, Erynnini, Pyrgini, Aeromachini, Baorini, Taractrocerini, Hesperiini, Pierini, Euchloeini, Hamearini, Riodinini, Miletini, Tarakini, Spalgini, Lycaenini, Theclini, Arhopalini, Zesiini, Amblypodiini, Catapaecilmatini, Loxurini, Horagini, Cheritrini, Iolaini, Remelanini, Hypolycaenini, Deudorigini, Eumaeini, Niphandini, Lycaenesthini, Polyommatini, Danaini, Euploeini, Prothoini, Charaxini, Faunidini, Elymniini, Zetherini, Melanitini, Satyrini, Limenitidini, Adoliadini, Argynnini, Heliconiini, Biblidini, Apaturini, Cyrestini, Pseudergolini, Melitaeini, Nymphalini, Junoniini, Kallimini, Acraeini, Cethosiini

PAPILIONOIDEA
PAPILIONINAE
Troidini
Losaria Moore, 1902
L. coon Fabricius, 1793 Common Clubtail
L. rhodifer Butler, 1876 Andaman Clubtail
Pachliopta Reakirt, 1865
P. aristolochiae Fabricius, 1775 Common Rose
P. hector Linnaeus, 1758 Crimson Rose
P. pandiyana Moore, 1881 Malabar Rose
Troides Huebner, 1819
T. aeacus C. & R. Felder, 1860 Golden Birdwing
T. helena Linnaeus, 1758 Common Birdwing
T. minos Cramer, 1779 Southern Birdwing
Atrophaneura Reakirt, 1865
A. aidoneus Doubleday, 1845 Lesser Batwing
A. varuna White, 1842 Common Batwing
Byasa Moore, 1882
B. crassipes Oberthur, 1893 Black Windmill
B. dasarada Moore, 1858 Great Windmill
B. latreillei Donovan, 1826 Rose Windmill
B. nevilli Wood-Mason, 1882 Nevill's Windmill
B. plutonius Oberthur, 1876 Chinese Windmill
B. polla de Niceville, 1897 de Niceville's Windmill
B. polyeuctes Doubleday, 1842 Common Windmill
Papilionini
Papilio Linnaeus, 1758
P. agestor Gray, 1831 Tawny Mime
P. alcmenor C. & R. Felder, 1864 Redbreast
P. arcturus Westwood, 1842 Blue Peacock
P. bianor Cramer, 1777 Common Peacock
P. bootes Westwood, 1842 Tailed Redbreast
P. buddha Westwood, 1872 Malabar Banded Peacock
P. castor Westwood, 1842 Common Raven
P. clytia Linnaeus, 1758 Common Mime
P. crino Fabricius, 1793 Common Banded Peacock
P. demoleus Linnaeus, 1758 Lime Butterfly
P. dravidarum Wood-Mason, 1880 Malabar Raven
P. elephenor Doubleday, 1845 Yellow-crested Spangle
P. epycides Hewitson, 1864 Lesser Mime
P. helenus Linnaeus, 1758 Red Helen
P. krishna Moore, 1858 Krishna Peacock
P. liomedon Moore, 1875 Malabar Banded Swallowtail
P. machaon Linnaeus, 1758 Common Yellow Swallowtail
P. mayo Atkinson, 1874 Andaman Mormon
P. memnon Linnaeus, 1758 Great Mormon
P. nephelus Boisduval, 1836 Yellow Helen
P. paradoxa Zincken, 1831 Great Blue Mime
P. paris Linnaeus, 1758 Paris Peacock
P. polymnestor Cramer, 1775 Blue Mormon
P. polytes Linnaeus, 1758 Common Mormon
P. prexaspes C. & R. Felder, 1865 Andaman Helen
P. protenor Cramer, 1775 Spangle
P. slateri Hewitson, 1859 Blue-striped Mime
P. xuthus Linnaeus, 1767 Chinese Yellow Swallowtail
Leptocircini
Graphium Scopoli, 1777
G. adonarensis Rothschild, 1896 Cryptic Bluebottle
G. agamemnon Linnaeus, 1758 Tailed Jay
Graphium agetes Westwood, 1843 Fourbar Swordtail
G. albociliatis Fruhstorfer, 1901 Scarce Jay
G. antiphates Cramer, 1775 Fivebar Swordtail
G. aristeus Stoll, 1780 Chain Swordtail
G. arycles Boisduval, 1836 Spotted Jay
G. chironides Honrath, 1884 Veined Jay
G. cloanthus Westwood. 1841 Glassy Bluebottle
G. doson C. & R. Felder, 1864 Common Jay
G. epaminondas Oberthur, 1789 Andaman Swordtail
G. eurous Leech, 1893 Sixbar Swordtail
G. eurypylus Linnaeus, 1758 Great Jay
G. macareus Godart, 1819 Lesser Zebra
G. mandarinus Oberthur, 1879 Spectacle Swordtail
G. megarus Westwood, 1844 Spotted Zebra
G. nomius Esper, 1799 Spot Swordtail
G. sarpedon Linnaeus, 1758 Common Bluebottle
G. teredon C. & R. Felder, 1865 Southern Bluebottle
G. xenocles Doubleday, 1842 Great Zebra
Lamproptera Gray, 1832
L. curius Fabricius, 1787 White Dragontail
L. meges Zinken, 1831 Green Dragontail
Teinopalpini
Meandrusa Moore, 1888
M. lachinus Fruhstorfer, 1902 Brown Gorgon
M. payeni Boisduval, 1836 Yellow Gorgon
Teinopalpus Hope, 1843
T. imperialis Hope, 1843 Kaiser-i-Hind

PARNASSIINAE
Zerynthiini
Bhutanitis Atkinson, 1873
B. lidderdalii Atkinson, 1873 Bhutan Glory
B. ludlowi Gabriel, 1942 Ludlow's Bhutan Glory
Parnassiini
Parnassius Latreille, 1804
P. Parnassius actius Eversmann, 1843 Scarce Red Apollo
P. Parnassius epaphus Oberthur, 1879 Common Red Apollo
P. Parnassius jacquemontii Boisduval, 1836 Keeled Apollo
P. Parnassius tianschanicus Oberthur, 1879 Large Keeled Apollo
P. Kailasus charltonius Gray, 1853 Regal Apollo
P. Kailasius loxias Pungeler, 1901 Stately Apollo
P. Kailasius imperator augustus Fruhstorfer, 1903 Noble Apollo
P. Kailasius acdestis Grum-Grshimailo, 1891 Dusky Apollo

P. Koramius staudingeri hunza Grum-Grshimailo, 1888
P. Koramius staudingeri mamaievi Bang-Haas, 1915
P. Koramius stenosemus Honrath, 1890 Greater Banded Apollo
P. Koramius stoliczkanus C. & R. Felder, 1865 Lesser Banded Apollo
P. Koramius kumaonensis Riley, 1926 Himalayan Banded Apollo
P. Tadumia acco Gray, 1853 Varnished Apollo
P. Tadumia maharaja Avinoff, 1916 Royal Apollo
P. Lingamius hardwickii Gray, 1831 Common Blue Apollo
P. Kreizbergia simo Gray, 1853 Black-edged Apollo

HESPERIIDAE
COELIADINAE
Badamia Moore, 1881
B. exclamationis Fabricius, 1775 Brown Awl
Bibasis Moore, 1881
B. iluska Hewitson, 1867 Slate Awlet
B. sena Moore, 1866 Orange-tailed Awlet
Burara Swinhoe, 1893
B. amara Moore, 1866 Small Green Awlet
B. anadi de Niceville, 1884 Plain Orange Awlet
B. etelka Hewitson, 1867 Great Orange Awlet
B. gomata Moore, 1866 Pale Green Awlet
B. harisa Moore, 1866 Orange Striped Awlet
B. jaina Moore, 1866 Orange Awlet
B. oedipodea Swainson, 1820 Branded Orange Awlet
B. vasutana Moore, 1866 Green Awlet
Hasora Moore, 1881
H. anura de Niceville, 1889 Slate Awl
H. badra Moore, 1858 Common Awl
H. chromus Cramer, 1780 Common Banded Awl
H. danda Evans, 1949 Purple Awl
H. khoda Mabille, 1876 Large Banded Awl
H. leucospila Mabille, 1891 Violet Awl
H. schoenherr Latreille, 1824 Yellow Banded Awl
H. taminatus Huebner, 1818 White Banded Awl
H. vitta Butler, 1870 Plain Banded Awl
Choaspes Moore, 1881
C. benjaminii Guerin-Meneville, 1843 Indian Awlking
C. furcatus Evans,1932 Hooked Awlking
C. stigmatus Evans, 1932 Branded Awlking
C. xanthopogon Kollar, 1844 Similar Awlking

EUDAMINAE
Lobocla Moore, 1884
L. liliana Atkinson, 1871 Marbled Flat

PYRGINAE
Tagiadini
Capila Moore, 1866
C. jayadeva Moore, 1866 Striped Dawnfly
C. lidderdali Elwes, 1888 Lidderdale's Dawnfly
C. pennicillatum de Niceville, 1893 Fringed Dawnfly
C. phanaeus Hewitson, 1867 Fulvous Dawnfly
C. pieridoides Moore, 1878 White Dawnfly
C. zennara Moore, 1866 Pale Striped Dawnfly
Tapena Moore, 1881
T. thwaitesi Moore, 1881 Black Angle
Darpa Moore, 1866
D. hanria Moore, 1866 Hairy Angle
D. pteria Hewitson, 1868 Snowy Angle
D. striata Druce, 1873 Striated Angle
Odina Mabille, 1891
O. decoratus Hewitson, 1867 Zigzag Flat
Coladenia Moore, 1881
C. agni de Niceville, 1884 Brown Pied Flat
C. agnioides Elwes & Edwards, 1897 Elwes's Pied Flat
C. indrani Moore, 1866 Tricolour Pied Flat
C. laxmi de Niceville, 1889 Grey Pied Flat
Satarupa Moore, 1866
S. gopala Moore, 1866 Large White Flat
S. splendens Tytler, 1915 Splendid White Flat
S. zulla Tytler, 1915 Tytler's White Flat
Seseria Matsumura, 1919
S. dohertyi Watson, 1893 Himalayan White Flat
S. sambara Moore, 1866 Sikkim White Flat
Pintara Evans, 1932
P. tabrica Hewitson, 1873 Crenulate Orange Flat
Chamunda Evans, 1949
C. chamunda Moore, 1866 Olive Flat
Gerosis Mabille, 1903
G. bhagava Moore, 1866 Common Yellow-breast Flat
G. phisara Moore, 1884 Dusky Yellow-breast Flat
G. sinica C. & R. Felder, 1862 White Yellow-breast Flat
Tagiades Huebner, 1819
T. calligana Butler, 1879 Malayan Snow Flat
T. cohaerens Mabille, 1914 Striped Snow Flat
T. gana Moore, 1866 Large Snow Flat
T. japetus Stoll, 1781 Suffused Snow Flat
T. menaka Moore, 1866 Spotted Snow Flat
T. parra Fruhstorfer, 1910 Common Snow Flat
T. litigiosa Moeschler, 1878 Water Snow Flat
Mooreana Evans, 1926
M. trichoneura C. & R. Felder, 1860 Yellow Flat
Ctenoptilum de Niceville, 1890
C. multiguttata de Niceville, 1890 Multispot Angle
C. vasava Moore, 1866 Tawny Angle
Odontoptilum de Niceville, 1890
O. angulata C. & R. Felder, 1862 Chestnut Angle
Caprona Wallengren, 1857
C. agama Moore, 1858 Spotted Angle
C. alida de Niceville, 1891 Evans' Angle
C. ransonnetii Felder, 1868 Golden Angle
Celaenorrhini
Celaenorrhinus Huebner, 1819
C. ambareesa Moore, 1866 Malabar Flat
C. andamanicus Wood Mason & de Niceville, 1881
C. asmara Butler, 1877 White Banded Flat
C. aspersa Leech, 1891 Large Streaked Flat
C. aurivittatus Moore, 1878 Dark Yellow-banded Fat
C. badius Hewitson, 1877 Scarce Banded Flat
C. dhanada Moore, 1866 Himalayan Yellow-banded Flat
C. ficulnea Hewitson, 1868 Velvet Flat
C. flavicincta de Niceville, 1887 Bhutan Flat
C. leucocera Kollar, 1844 Common Spotted Flat
C. morena Evans, 1949 Evans' Spotted Flat
C. munda Moore, 1884 Himalayan Spotted Flat
C. nigricans de Niceville, 1885 Small Banded Flat
C. patula de Niceville, 1889 Large Spotted Flat
C. pero de Niceville, 1889 Mussoorie Spotted Flat
C. plagifera de Niceville, 1889 de Niceville's Spotted Flat
C. pulomaya Moore, 1866 Multi-spotted Flat
C. putra Moore, 1866 Bengal Spotted Flat
C. pyrrha de Niceville, 1889 Double Spotted Flat
C. ratna Fruhstorfer, 1909 Tytler's Multi-spotted Flat
C. ruficornis Mabille, 1878 Tamil Spotted Flat
C. sumitra Moore, 1866 Moore's Spotted Flat
C. tibetanus Mabille, 1876 Tibet Flat
C. zea Swinhoe, 1909 Swinhoe's Flat
Pseudocoladenia Shirozu & Saigusa, 1962
P. dan Fabricius, 1787 Fulvous Pied Flat
P. fatua Evans, 1949 Sikkim Pied Flat

P. festa Evans, 1949 Naga Pied Flat
Sarangesa Moore, 1881
S. purendra Moore, 1882 Spotted Small Flat
S. dasahara Moore, 1866 Common Small Flat
Gomalia Moore, 1879
G. elma Trimen, 1862 African Marbled Skipper
Carcharodini
Carcharodus Huebner, 1819
C. alceae Esper, 1780 Plain Marbled Skipper
C. dravira Moore, 1875 Tufted Marbled Skipper
Spialia Swinhoe, 1912
S. doris Walker, 1870 Sind Skipper
S. galba Fabricius, 1793 Indian Skipper
S. orbifer Huebner, 1823 Brick Skipper
Erynnini
Erynnis Schrank, 1801
E. pathan Evans, 1949 Inky Skipper
Pyrgini
Pyrgus Huebner, 1819
P. alpinus Erschoff, 1874 Mountain Skipper
P. cashmirensis Moore, 1874 Kashmir Skipper

HETEROPTERINAE
Carterocephalus Lederer, 1852
C. avanti de Niceville, 1886 Orange-and-silver Hopper

HESPERIINAE
Aeromachini
Ochus de Niceville, 1894
O. subvittatus Moore, 1878 Tiger Hopper
Baracus Moore, 1881
B. vittatus Felder, 1862 Hedge Hopper
Ampittia Moore, 1881
A. dioscorides Fabricius, 1793 Bush Hopper
A. maroides de Niceville, 1896 Scarce Bush Hopper
Aeromachus de Niceville, 1890
A. dubius Elwes & Edwards, 1897 Dingy Scrub Hopper
A. jhora de Niceville, 1885 Grey Scrub Hopper
A. kali de Niceville, 1895 Blue-spotted Scrub Hopper
A. pygmaeus Fabricius, 1775 Pigmy Scrub Hopper
A. stigmatus Moore, 1878 Veined Scrub Hopper
Sebastonyma Watson, 1893
S. dolopia Hewitson, 1868 Tufted Ace
Sovia Evans, 1949
S. grahami Evans, 1926 Graham's Ace
S. hyrtacus de Niceville, 1897 White-branded Ace
S. lucasii Mabille, 1876 Lucas' Ace
S. malta Evans, 1949 Manipur Ace
S. separata Moore, 1882 Chequered Ace
Pedesta Hemming, 1934
P. masuriensis Moore, 1878 Mussoorie Bush Bob
P. panda Evans, 1937 Naga Bush Bob
P. pandita de Niceville, 1885 Brown Bush Bob
Thoressa Swinhoe, 1913
T. aina de Niceville, 1889 Garhwal Ace
T. astigmata Swinhoe, 1890 Southern Spotted Ace
T. cerata Hewitson, 1876 Northern Spotted Ace
T. evershedi Evans, 1910 Evershed's Ace
T. fusca Elwes, 1893 Fuscous Ace
T. gupta de Niceville, 1886 Olive Ace
T. honorei de Niceville, 1887 Madras Ace
T. hyrie de Niceville, 1891 Largespot Plain Ace
T. masoni Moore, 1878 Mason's Ace
T. sitala de Niceville, 1885 Tamil Ace
Halpe Moore, 1878
H. arcuata Evans, 1937 Evans' Ace
H. filda Evans, 1949 Elwes's Ace
H. flava Evans, 1926 Tavoy Sulphur Ace
H. hauxwelli Evans, 1937 Hauxwell's Ace
H. homolea Hewitson, 1868 Indian Ace
H. knyvetti Elwes & Edwards, 1897 Knyvett's Ace
H. kumara de Niceville, 1885 Plain Ace
H. kusala Fruhstorfer, 1911 Tenasserim Ace
H. porus Mabille, 1877 Moore's Ace
H. sikkima Moore,1882 Sikkim Ace
H. wantona Swinhoe, 1893 Confusing Ace
H. zema Hewitson, 1877 Banded Ace
H. zola Evans, 1937 Long-banded Ace
Pithauria Moore, 1878
P. marsena Hewitson, 1866 Branded Straw Ace
P. murdava Moore, 1866 Dark Straw Ace
P. stramineipennis Wood-Mason & de Niceville, 1887
Apostictopterus Leech, 1893
A. fuliginosus Leech, 1893 Giant Hopper
Astictopterus C. & R. Felder, 1860
A. jama C. & R. Felder, 1860 Forest Hopper
Arnetta Watson, 1893
A. atkinsoni Moore, 1878 Atkinson's Bob
A. mercara Evans, 1932 Coorg Forest Bob
A. vindhiana Moore, 1883 Vindhyan Bob
Actinor Watson, 1893
A. radians Moore, 1878 Veined Dart
Iambrix Watson, 1893
I. salsala Moore, 1866 Chestnut Bob
Koruthaialos Watson, 1893
K. butleri de Niceville, 1884 Dark Velvet Bob
K. rubecula Ploetz, 1882 Narrow-banded Velvet Bob
K. sindu C. & R. Felder, 1860 Bight Red Velvet Bob
Psolos Staudinger, 1889
P. fuligo Mabille, 1876 Coon
Stimula de Niceville, 1898
S. swinhoei Elwes & Edwards, 1897 Watson's Demon
Ancistroides Butler, 1874
A. nigrita Latreille, 1824 Chocolate Demon
Notocrypta de Niceville, 1889
N. curvifascia C. & R. Felder, 1862 Restricted Demon
N. feisthamelii Boisduval, 1832 Spotted Demon
N. paralysos Wood-Mason & de Niceville, 1881 Common Banded Demon
Udaspes Moore, 1881
U. folus Cramer, 1775 Grass Demon
Scobura Elwes & Edwards, 1897
S. cephala Hewitson, 1876 Forest Bob
S. cephaloides de Niceville, 1889 Large Forest Bob
S. isota Swinhoe, 1893 Swinhoe's Forest Bob
S. phiditia Hewitson, 1866 Malay Forest Bob
S. tytleri Evans, 1914 Tytler's Forest Bob
S. woolletti Riley, 1923 Wollett's Forest Bob
Suada de Niceville, 1895
S. swerga de Niceville, 1884 Grass
Suastus Moore, 1881
S. gremius Fabricius, 1798 Indian Palm Bob
S. minutus Moore, 1877 Ceylon Palm Bob
Cupitha Moore, 1884
C. purreea Moore, 1877 Wax Dart
Zographetus Watson, 1893
Z. ogygia Hewitson, 1866 Purple-spotted Flitter
Z. rama Mabille, 1877 Small Flitter
Z. satwa de Niceville, 1884 Purple-and-gold Flitter
Hyarotis Moore, 1881
H. adrastus Stoll, 1780 Tree Flitter
H. microstictum Wood-Mason & de Niceville, 1887 Brush Flitter
Quedara Swinhoe, 1919

Q. basiflava de Niceville, 1888 Golden Flitter
Q. monteithi Wood-Mason & de Niceville, 1887 Dubious Flitter
Isma Distant, 1886
I. bonota Cantlie & Norman, 1959 Assam Lancer
Plastingia Butler, 1870
P. naga de Niceville, 1884 Silver-spotted Lancer
Salanoemia Eliot, 1978
S. fuscicornis Elwes & Edwards, 1897 Purple Lancer
S. noemi de Niceville, 1885 Spotted Yellow Lancer
S. sala Hewitson, 1866 Maculate Lancer
S. tavoyana Evans, 1926 Yellow-streaked Lancer
Pyroneura Eliot, 1978
P. margherita Doherty, 1889 Assamese Yellow-vein Lancer
P. niasana Fruhstorfer, 1909 Red-vein Lancer
Lotongus Distant, 1886
L. sarala de Niceville, 1889 Yellowband Palmer
Zela de Niceville, 1895
Z. zeus de Niceville, 1895 Redeye Palmer
Gangara Moore, 1881
G. lebadea Hewitson, 1886 Banded Redeye
G. thyrsis Fabricius, 1775 Giant Redeye
Erionota Mabille, 1878
E. hiraca Moore, 1881 Andaman Redeye
E. thrax Linnaeus, 1767 Palm Redeye
E. torus Evans, 1941 Banana Skipper
Matapa Moore, 1881
M. aria Moore, 1866 Common Redeye
M. cresta Evans, 1949 Darkbrand Redeye
M. druna Moore, 1866 Greybrand Redeye
M. purpurascens Elwes & Edwards, 1897 Purple Redeye
M. sasivarna Moore, 1866 Black-veined Redeye
Pudicitia de Niceville, 1895
P. pholus de Niceville, 1889 Spotted Redeye
Unkana Distant, 1886
U. ambasa Moore, 1858 Hoary Palmer
Hidari Distant, 1886
H. bhawani de Niceville, 1889 Veined Palmer
Pirdana Distant, 1886
P. hyela Hewitson, 1867 Green-striped Palmer
P. distanti Staudinger, 1889 Plain Green Palmer
Creteus de Niceville, 1895
C. cyrina Hewitson, 1876 Nonsuch Palmer
Baorini
Gegenes Huebner, 1819
G. nostrodamus Fabricius, 1793 Dingy Swift
G. pumilio Hoffmansegg, 1804 Pygmy Swift
Parnara Moore, 1881
P. ganga Evans, 1937 Evans' Swift
P. guttatus Bremer & Grey, 1852 Straight Swift
P. bada Moore, 1878 Ceylon Swift
Borbo Evans, 1949
B. bevani Moore, 1878 Bevan's Swift
B. cinnara Wallace, 1866 Rice Swift
Pelopidas Walker, 1870
P. agna Moore, 1866 Obscure Branded Swift
P. assamensis de Niceville, 1882 Great Swift
P. conjuncta Herrich-Schaeffer, 1869 Conjoined Swift
P. mathias Fabricius, 1798 Variable Swift
P. sinensis Mabille, 1877 Large Branded Swift
P. subochracea Moore, 1878 Moore's Swift
P. thrax Huebner, 1821 Small Branded Swift
Polytremis Mabille, 1904
P. discreta Elwes & Edwards, 1897 Himalayan Swift
P. eltola Hewitson, 1869 Yellow Spot Swift
P. lubricans Herrich-Schaeffer, 1869 Contiguous Swift
P. minuta Evans, 1926 Baby Swift
Baoris Moore, 1881
B. chapmani Evans, 1937 Small Paintbrush Swift
B. farri Moore, 1878 Paintbrush Swift
B. pagana de Niceville, 1887 Figure of Eight Swift
B. unicolor Moore, 1883 Black Paintbrush Swift
Caltoris Swinhoe, 1893
C. aurociliata Elwes & Edwards, 1897 Yellow Fringed Swift
C. bromus Leech, 1894 Leech's Swift
C. brunnea Snellen, 1876 Dark-branded Swift
C. cahira Moore, 1877 Colon Swift
C. canaraica Moore, 1884 Canara Swift
C. confusa Evans, 1932 Confusing Swift
C. cormasa Hewitson, 1876 Full-stop Swift
C. kumara Moore, 1878 Blank Swift
C. philippina Herrich-Schaeffer, 1869 Philippine Swift
C. plebeia de Niceville, 1887 Tufted Swift
C. sirius Evans, 1926 Sirius Swift
C. tulsi de Niceville, 1884 Purple Swift
Iton de Niceville 1895
I. semamora Moore, 1866 Common Wight
Taractrocerini
Taractocera Butler, 1870
T. ceramas Hewitson, 1868 Tamil Grassdart
T. danna Moore, 1865 Himalayan Grass Dart
T. maevius Fabricius, 1793 Common Grass Dart
Oriens Evans, 1932
O. concinna Elwes & Edwards, 1897 Tamil Dartlet
O. gola Moore, 1877 Common Dartlet
O. goloides Moore, 1881 Ceylon Dartlet
O. paragola de Niceville, 1895 Malay Dartlet
Potanthus Scudder, 1872
P. confucius C. & R. Felder, 1862 Chinese Dart
P. dara Kollar, 1844 Himalayan Dart
P. ganda Fruhstorfer, 1911 Sumatran Dart
P. flavus Murray, 1875 Japanese Dart
P. hetaerus Mabille, 1883 Large Dart
P. juno Evans, 1932 Burmese Dart
P. lydia Evans, 1934 Forest Dart
P. mara Evans, 1932 Sikkim Dart
P. mingo Edwards, 1866 Narrow Bi-dent Dart
P. nesta Evans, 1934 Brandless Dart
P. pallidus Evans, 1932 Pale Dart
P. palnia Evans, 1914 Palni Dart
P. pava Fruhstorfer, 1911 Yellow Dart
P. pseudomaesa Moore, 1881 Indian Dart
P. rectifasciata Elwes & Edwards, 1897 Branded Dart
P. sita Evans, 1932 Yellow-and-black Dart
P. trachala Mabille, 1878 Broad Bi-dent Dart
Telicota Moore, 1881
T. augias Linnaeus, 1763 Pale Palm Dart
T. bambusae Moore, 1878 Dark Palm Dart
T. besta Evans, 1949 Large Palm Dart
T. colon Fabricius, 1775 Common Palm Dart
T. linna Evans, 1949 Evans' Palm Dart
T. ohara Ploetz, 1883 Crested Palm Dart
Cephrenes Waterhouse & Lyell, 1914
C. acalle Hopffer, 1874 Plain Palm Dart
Hesperiini
Hesperia Fabricius, 1793
H. comma Linnaeus, 1758 Chequered Darter
Ochlodes Scudder, 1872
O. brahma Moore, 1878 Himalayan Darter
O. siva Moore, 1878 Assam Darter
O. subhyalina Bremer & Grey, 1853 Sub-hyaline Darter

PIERIDAE
COLIADINAE
Catopsilia Huebner, 1819
C. pomona Fabricius, 1775 Common Emigrant

C. pyranthe Linnaeus, 1758 Mottled Emigrant
Dercas Doubleday, 1847
D. verhuelli van der Hoeven, 1839 Tailed Sulphur
D. lycorias Doubleday, 1842 Plain Sulphur
Gonepteryx Leach, 1815
G. amintha Blanchard, 1871 Chinese Brimstone
G. chitralensis Moore, 1905 Karakoram Brimstone
G. mahaguru Gistel, 1857 Lesser Brimstone
G. nepalensis Doubleday, 1847 Himalayan Brimstone
Gandaca Moore, 1906
G. harina Horsfield, 1829 Tree Yellow
Eurema Huebner, 1819
E. andersoni Moore, 1886 One-spot Grass Yellow
E. blanda Boisduval, 1836 Three-spot Grass Yellow
E. brigitta Stoll, 1780 Small Grass Yellow
E. hecabe Linnaeus, 1758 Common Grass Yellow
E. laeta Boisduval, 1836 Spotless Grass Yellow
E. nilgiriensis Yata, 1990 Nilgiri Grass Yellow
E. simulatrix Semper, 1891 Scarce Grass Yellow
Colias Fabricius, 1807
C. cocandica Erschoff, 1874 Pamir Clouded Yellow
C. dubia Elwes, 1906 Dwarf Clouded Yellow
C. eogene C. & R. Felder, 1865 Fiery Clouded Yellow
C. erate Esper, 1805 Pale Clouded Yellow
C. fieldi Menetries, 1855 Dark Clouded Yellow
C. ladakensis C. & R. Felder, 1865 Ladak Clouded Yellow
C. leechi Grum-Grshimailo, 1893 Glaucous Clouded Yellow
C. marcopolo Grum-Grshimailo, 1888 Marcopolo's Clouded Yellow
C. nilagiriensis C. & R. Felder, 1859 Nilgiri Clouded Yellow
C. stoliczkana Moore, 1878 Orange Clouded Yellow
C. thrasibulus Fruhstorfer, 1910 Lemon Clouded Yellow
C. wiskotti Staudinger, 1882 Broad-bordered Clouded Yellow

PIERINAE
Pierini
Leptosia Huebner, 1818
L. nina Fabricius, 1793 Psyche
Baltia Moore, 1878
B. butleri Moore, 1882 Butler's Dwarf
B.shawii Bates, 1873 Shaw's Dwarf
B.sikkima Fruhstorfer, 1903 Sikkim Dwarf
Mesapia Gray, 1856
M. peloria Hewitson, 1853 Thibet Blackvein
Aporia Huebner, 1819
A. agathon Gray, 1831 Great Blackvein
A. harrietae de Niceville, 1893 Bhutan Blackvein
A. leucodice Eversmann, 1843 Baluchi Blackvein
A. nabellica Boisduval, 1836 Dusky Blackvein
A. soracta Moore, 1857 Himalayan Blackvein
Pieris Schrank, 1801
P. ajaka Moore, 1865 Himalayan White
P. brassicae Linnaeus, 1758 Large Cabbage White
P. canidia Linnaeus, 1768 Indian Cabbage White
P. deota de Niceville, 1884 Kashmir White
P. dubernardi Oberthur, 1884 Chumbi White
P. extensa Poujade, 1888 Bhutan White
P. krueperi Staudinger, 1860 Green Banded White
P. melete Menetries, 1857 Western Black-Veined White
P. rapae Linnaeus, 1758 Small Cabbage White
Talbotia Bernardi, 1958
T. naganum Moore, 1884 Naga White
Pontia Fabricius, 1807
P. callidice Huebner, 1799-1800 Lofty Bath White
P. chloridice Huebner, 1808-13 Lesser Bath White
P. daplidice Linnaeus, 1758 Bath White
P. glauconome Klug, 1829 Desert Bath White
P. sherpae Epstein, 1979 Sherpa White
Ixias Huebner, 1819
I. marianne Cramer, 1779 White Orange Tip
I. pyrene Linnaeus, 1764 Yellow Orange Tip
Colotis Huebner, 1819
C. amata Cramer, 1775 Small Salmon Arab
C. aurora Cramer, 1780 Plain Orange Tip
C. danae Fabricius, 1775 Crimson Tip
C. etrida Boisduval, 1836 Little Orange Tip
C. fausta Olivier, 1804 Large Salmon Arab
C. phisadia Godart, 1819 White Arab
C. protractus Butler, 1876 Blue Spotted Arab
Appias Huebner, 1819
A. albina Boisduval, 1836 Common Albatross
A. indra Moore, 1857 Plain Puffin
A. lalage Doubleday, 1842 Spot Puffin
A. libythea Fabricius, 1775 Striped Albatross
A. lyncida Cramer, 1777 Chocolate Albatross
A. nero Fabricius, 1793 Orange Albatross
A. paulina Cramer, 1777 Lesser Albatross
A. wardi Moore, 1884 Indian Albatross
Saletara Distant, 1885
S. liberia Cramer, 1779 Nicobar Albatross
Prioneris Wallace, 1867
P. philonome Boisduval, 1836 Redspot Sawtooth
P. sita C. & R. Felder, 1865 Painted Sawtooth
P. thestylis Doubleday, 1842 Spotted Sawtooth
Belenois Huebner, 1819
B. aurota Fabricius, 1793 Pioneer
Cepora Billberg, 1820
C. nadina Lucas, 1852 Lesser Gull
C. nerissa Fabricius, 1775 Common Gull
Delias Huebner, 1819
D. acalis Godart, 1819 Redbreast Jezabel
D. agostina Hewitson, 1852 Yellow Jezabel
D. belladonna Fabricius, 1793 Hill Jezabel
D. berinda Moore, 1872 Dark Jezabel
D. descombesi Boisduval, 1836 Redspot Jezabel
D. eucharis Drury, 1773 Common Jezabel
D. hyparete Linnaeus, 1758 Painted Jezabel
D. lativitta Leech, 1893 Broadwing Jezabel
D. sanaca Moore, 1857 Pale Jezabel
D. pasithoe Linnaeus, 1767 Redbase Jezabel
Euchloeini
Euchloe Huebner, 1819
E. daphalis Moore, 1865 Little White
Pareronia Bingham, 1907
P. avatar Moore, 1858 Pale Wanderer
P. ceylanica C. & R. Felder, 1865 Dark Wanderer
P. valeria Cramer, 1776 Common Wanderer
Hebomoia Huebner, 1819
H. glaucippe Linnaeus, 1758 Great Orange Tip

RIODININAE
Hamearini
Zemeros Boisduval, 1836
Z. flegyas Cramer 1780 Punchinello
Dodona Hewitson, 1861
D. adonira Hewitson, 1866 Striped Punch
D. dipoea Hewitson, 1866 Lesser Punch
D. durga Kollar, 1844 Common Punch
D. egeon Westwood, 1851 Orange Punch
D. eugenes Bates, 1868 Tailed Punch
D. longicaudata Niceville, 1881 Long-tailed Punch
D. ouida Moore, 1866 Mixed Punch
Riodinini
Abisara C. & R. Felder, 1860

A. *abnormis* Moore, 1884 Abnormal Judy
A. *attenuata* Tytler, 1915 Short Tailed Judy
A. *bifasciata* Moore, 1877 Twospot Plum Judy
A. *burnii* de Niceville, 1895 Burn's Judy
A. *chela* de Niceville, 1886 Spot Judy
A. *echerius* Stoll, 1790 Plum Judy
A. *fylla* Westwood, 1851 Dark Judy
A. *neophron* Hewitson, 1861 Tailed Judy
A. *saturata* Moore, 1878 Malayan Plum Judy
Taxila Doubleday, 1847
T. *haquinus* Fabricius, 1793 Harlequin
Stiboges Butler, 1876
S. *nymphidia* Butler, 1876 Columbine

LYCAENIDAE
CURETINAE
Curetis Huebner, 1819
C. *acuta* Moore, 1877 Angled Sunbeam
C. *bulis* Westwood, 1852 Bright Sunbeam
C. *naga* Evans, 1954 Naga Sunbeam
C. *saronis* Moore, 1877 Burmese Sunbeam
C. *siva* Evans, 1954 Siva Sunbeam
C. *thetis* Drury, 1773 Indian Sunbeam

PORITINAE
Poritia Moore, 1866
P. *erycinoides* C. & R. Felder, 1865 Blue Gem
P. *hewitsoni* Moore, 1866 Common Gem
Simiskina Distant, 1886
S. *phalena* Hewitson, 1874 Broad-banded Brilliant
Liphyrinae
Liphyra Westwood, 1864
L. *brassolis* Westwood, 1864 Moth Butterfly

MILETINAE
Miletini
Allotinus C. & R. Felder, 1865
A. *drumila* Moore, 1866 Great Darkie
A. *subviolaceus* C. & R. Felder, 1865 Blue Darkie
A. *unicolor* C. & R. Felder, 1865 Common Darkie
A. *taras* Doherty, 1889 Brown-tipped Darkie
Logania Distant, 1884
L. *distanti* Semper, 1889 Dark Mottle
L. *marmorata* Moore, 1884 Pale Mottle
L. *watsoniana* de Niceville, 1898 Watson's Mottle
Miletus Huebner, 1819
M. *chinensis* C. Felder, 1862 Common Brownie
M. *symethus* Cramer, 1777 Great Brownie
Tarakini
Taraka de Niceville, 1890
T. *hamada* Druce, 1875 Forest Pierrot
Spalgini
Spalgis Moore, 1879
S. *baiongus* Cantlie & Norman 1960 Assam Apefly
S. *epius* Westwood, 1852 Common Apefly

LYCAENINAE
Lycaenini
Lycaena Fabricius, 1807
L. *kasyapa* Moore, 1865 Green Copper
L. *panava* Westwood, 1852 White-bordered Copper
L. *phlaeas* Linnaeus, 1761 Common Copper
Thersamonia Verity, 1919
T. *aditya* Moore, 1875 Ladakh Copper
Heliophorus Geyer in Huebner, 1832
H. *androcles* Westwood, 1851 Green Sapphire
H. *bakeri* Evans, 1927 Western Blue Sapphire
H. *brahma* Moore, 1858 Golden Sapphire
H. *epicles* Godart, 1824 Purple Sapphire
H. *hybrida* Tytler, 1912 Hybrid Sapphire
H. *ila* de Niceville & Martin, 1896 Restricted Purple Sapphire
H. *indicus* Fruhstorfer,1908 Indian Purple Sapphire
H. *kohimensis* Tytler, 1912 Naga Sapphire
H. *moorei* Hewitson, 1865 Azure Sapphire
H. *oda* Hewitson, 1865 Eastern Blue Sapphire
H. *sena* Kollar, 1844 Sorrel Sapphire
H. *tamu* Kollar, 1844 Powdery Green Sapphire

APHNAEINAE
Apharitis Riley, 1925
A. *acamas* Klug, 1834 Tawny Silverline
A. *lilacinus* Moore, 1884 Lilac Silverline
Spindasis Wallengren, 1857
S. *abnormis* Moore, 1884 Abnormal Silverline
S. *elima* Moore, 1877 Scarce Shot Silverline
S. *elwesi* Evans, 1925 Elwes' Silverline
S. *evansii* Tytler, 1915 Cinnamon Silverline
S. *ictis* Hewitson, 1865 Common Shot Silverline
S. *lohita* Horsfield, 1829 Long-banded Silverline
S. *nipalicus* Moore, 1884 Silvergrey Silverline
S. *rukma* de Niceville, 1889 Obscure Silverline
S. *rukmini* de Niceville, 1889 Khaki Silverline
S. *schistacea* Moore, 1881 Plumbeous Silverline
S. *syama* Horsfield, 1829 Club Silverline
S. *vulcanus* Fabricius, 1775 Common Silverline

THECLINAE
Theclini
Chaetoprocta de Niceville, 1890
C. *kurumi* Fujioka, 1970 Nepal Walnut Blue
C. *odata* Hewitson, 1865 Walnut Blue
Euaspa Moore, 1884
E *milionia* Hewitson, 1869 Water Hairstreak
E. *miyashitai* Koiwaya, 2002 Darjeeling Hairstreak
E. *mikamii* Koiwaya, 2002 Arunachal Hairstreak
E. *pavo* de Niceville, 1887 Peacock Hairstreak
Shizuyaozephyrus Koiwaya, 2003
S. *ziha* Hewitson, 1865 White-spotted Hairstreak
Fujiokaozephyrus Koiwaya, 2007
F.tsangkie Oberthur, 1886 Suroifui Hairstreak
Iwaseozephyrus Fujioka 1994
I. *mandara* Doherty, 1886 Indian Purple Hairstreak
Esakiozephyrus Shirozu & Yamamoto, 1956
E. *icana* Moore, 1875 Dull Green Hairstreak
Chrysozephyrus Shirozu & Yamamoto, 1956
C. *disparatus* Howarth, 1957 Howarth's Green Hairstreak
C. *duma* Hewitson, 1878 Metallic Green Hairstreak
C. *dumoides* Tytler, 1915 Broad-bordered Green Hairstreak
C. *intermedius* Tytler, 1915 Intermediate Green Hairstreak
C. *kabrua* Tytler, 1915 Kabru Green Hairstreak
C. *sandersi* Howarth, 1957 Sanders' Green Hairstreak
C. *sikkimensis* Howarth, 1957 Sikkim Green Hairstreak
C. *tytleri* Howarth, 1957 Manipur Green Hairstreak
C. *vittatus* Tytler, 1915 Tytler's Green Hairstreak
C. *zoa* de Niceville, 1889 Powdered Green Hairstreak
Neozephyrus Sibatani et Ito, 1942
N. *suroia* Tytler, 1915 Cerulean Hairstreak
Shirozuozephyrus Koiwaya, 2007
S. *bhutanensis* Howarth, 1957 Bhutan Hairstreak
S. *birupa* Moore, 1877 Fawn Hairstreak
S. *jakamensis* Tytler, 1915 Jakama Hairstreak
S. *khasia* de Niceville, 1890 Tailless Metallic Green Hairstreak
S. *kirbariensis* Tytler, 1915 Kirbari Hairstreak
S. *paona* Tytler, 1915 Paona Hairstreak

S. triloka Hannyngton, 1910 Kumaon Hairstreak
Inomataozephyrus Koiwaya, 2007
I. assamicus Tytler, 1915 Assam Hairstreak
I. syla Kollar, 1844 Silver Hairstreak
Thermozephyrus Inomata et Itagaki, 1986
T. ataxus Westwood, 1851 Wonderful Hairstreak
Leucantigius Shirozu et Murayama, 1951
L. atayalicus Shirozu et Murayama, 1943 Pale Hairstreak
Amblopala Leech, 1893
A. avidiena Hewitson, 1877 Chinese Hairstreak
Arhopalini
Arhopala Boisduval, 1832
A. aberrans de Niceville, 1889 Pale Bushblue
A. abseus Hewitson, 1862 Aberrant Bushblue
A. ace de Niceville, 1893 Tytler's Dull Oakblue
A. aeeta de Niceville, 1893 Dawnas Tailless Oakblue
A. alea Hewitson, 1862 Kanara Oakblue
A. allata Staudinger, 1889 Tytler's Rosy Oakblue
A. agrata de Niceville, 1890 de Niceville's Dull Oakblue
A. alax Evans, 1932 Silky Oakblue
A. alesia C. & R. Felder, 1865 Pallid Oakblue
A. amantes Hewitson, 1862 Large Oakblue
A. ammonides Doherty, 1891 Dark Bushblue
A. anarte Hewitson, 1862 Magnificient Oakblue
A. ariel Doherty, 1891 Chocolate Bushblue
A. arvina Hewitson, 1863 Purple-brown Tailless Oakblue
A. asinarus C. & R. Felder, 1865 Broad-banded Oakblue
A. asopia Hewitson, 1869 Plain Tailless Oakblue
A. athada Staudinger, 1889 Vinous Oakblue
A. atrax Hewitson, 1862 Indian Oakblue
A. aurelia Evans, 1925 Grey-washed Oakblue
A. bazaloides Hewitson, 1878 Tamil Oakblue
A. bazalus Hewitson, 1862 Powdered Oakblue
A. belphoebe Doherty, 1889 Doherty's Oakblue
A. birmana Moore, 1884 Burmese Bushblue
A. camdeo Moore, 1858 Lilac Oakblue
A. centaurus Fabricius, 1775 Centaur Oakblue
Arhopala comica de Niceville, 1900 Comic Oakblue
A. curiosa Evans, 1957 Bhutan Oakblue
A. democritus Fabricius, 1773 White-spotted Oakblue
A. dispar Riley & Godfrey, 1921 Frosted Oakblue
A. dodonea Moore, 1858 Pale Himalayan Oakblue
A. eumolphus Cramer, 1780 Green Oakblue
A. fulla Hewitson, 1862 Spotless Oakblue
A. ganesa Moore, 1858 Tailless Bushblue
A. hellenore Doherty, 1889 Pointed Green Oakblue
A. khamti Doherty, 1891 Doherty's Dull Oakblue
A. nicevillei Bethune-Baker, 1903 Large Spotted Oakblue
A. oenea Hewitson, 1869 Hewitson's Dull Oakblue
A. opalina Moore, 1884 Opal Oakblue
A. paraganesa de Niceville, 1882 Dusky Bushblue
A. paralea Evans, 1925 Glazed Oakblue
A. paramuta de Niceville, 1884 Hooked Oakblue
A. perimuta Moore, 1858 Yellowdisc Tailless Oakblue
A. rama Kollar, 1844 Dark Himalayan Oakblue
A. ralanda Corbet, 1941 Bright Oakblue
A. selta Hewitson, 1869 Rosy Oakblue
A. silhetensis Hewitson, 1862 Sylhet Oakblue
A. singla de Niceville, 1885 Yellow-disc Oakblue
A. zeta Moore, 1877 Andaman Tailless Oakblue
Thaduka Moore, 1879
T. multicaudata Moore, 1879 Many-tailed Oakblue
Apporasa Moore, 1884
A. atkinsoni Hewitson, 1869 Crenulate Oakblue
Mahathala Moore, 1878
M. ameria Hewitson, 1862 Falcate Oakblue
Flos Doherty, 1889
F. apidanus Cramer, 1773 Plain Plushblue
F. adriana de Niceville, 1883 Vareigated Plushblue
F. areste Hewitson, 1862 Tailless Plushblue
F. asoka de Niceville, 1883 Spangled Plushblue
F. chinensis C. & R. Felder, 1865 Chinese Plushblue
F. diardi Hewitson, 1862 Bifid Plushblue
F. fulgida Hewitson, 1863 Shining Plushblue
Mota de Niceville, 1890
M. massyla Hewitson, 1869 Saffron
Surendra Moore, 1879
S. quercetorum Moore, 1858 Common Acacia Blue
S. vivarna Horsfield, 1829 Burmese Acacia Blue
Zinaspa de Niceville, 1890
Z. todara Moore, 1884 Silver Streaked Acacia Blue
Zesiini
Zesius Huebner, 1819
Z. chrysomallus Huebner, 1819 Redspot
Amblypodiini
Amblypodia Horsfield, 1829
A. anita Hewitson, 1862 Purple Leaf Blue
Iraota Moore, 1881
I. rochana Horsfield, 1829 Scarce Silverstreak Blue
I. timoleon Stoll, 1790 Silverstreak Blue
Catapaecilmatini
Catapaecilma Butler, 1879
C. major Druce, 1895 Common Tinsel
C. subochracea Elwes, 1893 Yellow Tinsel
Acupicta Eliot, 1973
A. delicatum de Niceville, 1887 Dark Tinsel
Loxurini
Loxura Horsfield, 1829
L. atymnus Stoll, 1780 Yamfly
Yasoda Doherty, 1889
Y. tripunctata Hewitson, 1863 Branded Yamfly
Drina de Niceville, 1890
D. donina Hewitson, 1865 Brown Yam
Horagini
Horaga Moore, 1881
H. albimacula Wood-Mason & de Niceville, 1881 Violet Onyx
H. onyx Moore, 1858 Common Onyx
H. syrinx C. Felder, 1860 Yellow Onyx
H. viola Moore, 1882 Brown Onyx
Rathinda Moore, 1881
R. amor Fabricius, 1775 Monkey Puzzle
Cheritrini
Cheritra Moore, 1881
C. freja Fabricius, 1793 Common Imperial
Cheritrella de Niceville, 1887
C. truncipennis de Niceville, 1887 Truncate Imperial
Ticherra de Niceville, 1887
T. acte Moore, 1858 Blue Imperial
Drupadia Moore, 1884
D. scaeva Hewitson, 1863 Blue Posy
Iolaini
Pratapa Moore, 1881
P. deva Moore, 1858
P. icetas Hewitson, 1865 Dark Blue Royal
P. icetoides Elwes, 1891 Blue Royal
Tajuria Moore, 1881
T. albiplaga de Niceville, 1887 Pallid Royal
T. cippus Fabricius, 1798 Peacock Royal
T. culta de Niceville 1896 Black-branded Royal
Tajuria deudorix Hewitson, 1869 Flash Royal
T. diaeus Hewitson, 1865 Straightline Royal
T. illurgioides de Niceville, 1890 Scarce White Royal
T. illurgis Hewitson, 1869 White Royal

T. isaeus Hewitson, 1865 Bornean Royal
T. ister Hewitson, 1865 Uncertain Royal
T. jehana Moore, 1883 Plains Blue Royal
T. luculenta Leech, 1890 Chinese Royal
T. maculata Hewitson, 1865 Spotted Royal
T. megistia Hewitson,1869 Orange-and-black Royal
T. melastigma de Niceville, 1884 Branded Royal
T. yajna Doherty, 1886 Chestnut-and-black Royal
Dacalana Moore, 1884
D. cotys Hewitson, 1865 White-banded Royal
D. penicilligera de Niceville, 1890 Double-tufted Royal
Maneca de Niceville, 1890
M. bhotea Moore, 1884 Slate Royal
Creon de Niceville, 1896
C. cleobis Godart, 1824 Broadtail Royal
Bullis de Niceville, 1897
B. buto de Niceville, 1895 Baby Royal
Eliotiana Koçak, 1996
E. jalindra Horsfield, 1829 Banded Royal
Neocheritra Distant, 1884
N. fabronia Hewitson, 1878 Pale Grand Imperial
Charana de Niceville, 1890
C. cepheis de Niceville, 1894
C. mandarina Hewitson, 1863 Mandarin Blue
Suasa de Niceville, 1890
S. lisides Hewitson, 1863 Red Imperial
Britomartis de Niceville, 1895
B. cleoboides Elwes, 1892 Azure Royal
Remelanini
Remelana Moore, 1884
R. jangala Horsfield, 1829 Chocolate Royal
Ancema Eliot, 1973
A. blanka de Niceville, 1894 Silver Royal
A. ctesia Hewitson, 1865 Bi-Spot Royal
Hypolycaenini
Hypolycaena C. & R. Felder, 1862
H. erylus Godart, 1824 Common Tit
H. narada Kunte, 2015 Banded Tit
H. nilgirica Moore, 1884 Nilgiri Tit
H. thecloides C. & R. Felder, 1860 Brown Tit
Chliaria Moore, 1884
C. kina Hewitson, 1869 Blue Tit
C. othona Hewitson, 1865 Orchid Tit
Zeltus de Niceville, 1890
Z. amasa Hewitson, 1865 Fluffy Tit
Deudorigini
Deudorix Hewitson, 1863
D. epijarbas Moore, 1857 Cornelian
D. gaetulia de Niceville, 1893 Assam Cornelian
Virachola Moore, 1881
V. dohertyi Tytler, 1915 Doherty's Guava Blue
V. isocrates Fabricius, 1793 Common Guava Blue
V. kessuma Horsfield, 1829 Whiteline Flash
V. perse Hewitson, 1863 Large Guava Blue
V. similis Hewitson, 1863 Scarce Guava Blue
Artipe Boisduval, 1870
A. eryx Linnaeus, 1771 Green Flash
Sinthusa Moore, 1884
S. chandrana Moore, 1882 Broad Spark
S. nasaka Horsfield, 1829 Narrow Spark
S. virgo Elwes, 1887 Pale Spark
Araotes Doherty, 1889
A. lapithis Moore, 1858 Witch
Bindahara Moore, 1881
B. phocides Fabricius, 1793 Plane
Rapala Moore, 1881
R. buxaria de Niceville, 1889 Shot Flash
R. damona Swinhoe, 1890 Malay Red Flash
R. dieneces Hewitson, 1878 Scarlet Flash
R. extensa Evans, 1926 Chitral Flash
R. iarbus Fabricius, 1787 Common Red Flash
R. lankana Moore, 1879 Malabar Flash
R. manea Hewitson, 1863 Slate Flash
R. nissa Kollar, 1844 Common Flash
R. pheretima Hewitson, 1863 Copper Flash
R. rectivitta Moore, 1879 Scarce Shot Flash
R. refulgens de Niceville, 1891 Refulgent Flash
R. melida Fruhstorfer, 1912 Brilliant Flash
R. rosacea de Niceville, 1889 Rosy Flash
R. rubida Tytler, 1926 Tytler's Flash
R. selira Moore, 1874 Himalayan Red Flash
R. scintilla de Niceville, 1890 Scarce Slate Flash
R. suffusa Moore 1878 Suffused Flash
R. tara de Niceville, 1889 Assam Flash
R. varuna Horsfield, 1829 Indigo Flash
Pamela Hemming, 1935
P. dudgeoni de Niceville, 1894 Lister's Hairstreak
Eumaeini
Ahlbergia Bryk, 1946
A. leechii de Niceville, 1893 Ferruginous Hairstreak
Strymon Huebner, 1818
S. mackwoodi Evans, 1914 Mackwood's Hairstreak
Superflua Strand, 1910
S. deria Moore, 1865 Moore's Hairstreak

POLYOMMATINAE
Niphandini
Niphanda Moore, 1875
N. asialis de Niceville, 1895 Fawcett's Pierrot
N. cymbia de Niceville, 1884 Pointed Pierrot
Lycaenesthini
Anthene Doubleday, 1847
A. emolus Godart, 1824 Ciliate Blue
A. lycaenina Felder, 1868 Pointed Ciliate Blue
Polyommatini
Una de Niceville, 1890
U. usta Distant, 1886 Una
Orthomiella de Niceville, 1890
O. pontis Elwes, 1887 Straightwing Blue
O. rantaizana Wileman, 1910 Burmese Straightwing Blue
Petrelaea Toxopeus, 1929 Dingy Lineblue
P. dana de Niceville, 1884
Nacaduba Moore, 1881
N. berenice Herrich-Schäffer, 1869 Rounded Sixlineblue
N. beroe C. & R. Felder, 1865 Opaque Sixlineblue
N. calauria C. Felder, 1860 Dark Ceylon Sixlineblue
N. hermus C. Felder, 1865 Pale Fourlineblue
N. kurava Moore, 1858 Transparent Sixlineblue
N. pactolus C. Felder, 1860 Large Fourlineblue
N. pavana Horsfield, 1828 Small Fourlineblue
N. sanaya Fruhstorfer, 1916 Jewel Fourlineblue
N. subperusia Snellen, 1896 Violet Fourlineblue
Prosotas Druce, 1891
P. aluta Druce, 1873 Barred Lineblue
P. bhutea de Niceville, 1884 Bhutya Lineblue
P. dubiosa Semper, 1879 Tailless Lineblue
P. lutea Martin, 1895 Banded Lineblue
P. nora C. Felder, 1860 Common Lineblue
P. noreia R. Felder, 1868 White-tipped Lineblue
P. pia Toxopeus, 1929 Margined Lineblue
Ionolyce Toxopeus, 1829
I. helicon C. Felder, 1860 Pointed Lineblue
Catopyrops Toxopeus, 1929
C. ancyra C. Felder, 1860 Felder's Lineblue

Caleta Fruhstorfer, 1922
C. decidia Hewitson, 1876 Angled Pierrot
C. elna Hewitson, 1876 Elbowed Pierrot
C. roxus Godart, 1824 Straight Pierrot
Discolampa Toxopeus, 1929
D. ethion Westwood, 1851 Banded Blue Pierrot
Jamides Huebner, 1819
J. alecto C. Felder, 1860 Metallic Cerulean
J. bochus Stoll, 1782 Dark Cerulean
J. caeruleus H. Druce, 1873 Royal Cerulean
J. celeno Cramer, 1775 Common Cerulean
J. elpis Godart, 1924 Glistening Cerulean
J. ferrari Evans, 1932 Ferrar's Cerulean
J. kankena C. Felder, 1862 Frosted Cerulean
J. pura Moore, 1886 White Cerulean
Catochrysops Boisduval, 1832
C. panormus C. Felder, 1860 Silver Forgetmenot
C. strabo Fabricius, 1793 Forgetmenot
Lampides Huebner, 1819 Peablue
L. boeticus Linnaeus, 1767
Leptotes Scudder, 1876
L. plinius Fabricius, 1793 Zebra Blue
Castalius Huebner, 1819
C. rosimon Fabricius, 1775 Common Pierrot
Tarucus Moore, 1881
T. ananda de Niceville, 1884 Dark Pierrot
T. balkanicus Freyer, 1844 Black-spotted Pierrot
T. callinara Butler, 1886 Spotted Pierrot
T. hazara Evans, 1932 Hazara Pierrot
T. indicus Evans, 1932 Indian Pointed Pierrot
T. nara Kollar, 1848 Striped Pierrot
T. venosus Moore, 1882 Himalayan Pierrot
T. waterstradti Druce, 1895 Assam Pierrot
Zizeeria Chapman, 1910
Z. karsandra Moore, 1865 Dark Grass Blue
Pseudozizeeria Beuret, 1955
P. maha Kollar, 1844 Pale Grass Blue
Zizina Chapman 1910
Z. otis Fabricius, 1787 Lesser Grass Blue
Zizula Chapman, 1910
Z. hylax Fabricius, 1775 Tiny Grass Blue
Everes Huebner, 1819
E. argiades Pallas, 1771 Tailed Cupid
E. hugelii Gistel, 1857 Dusky-blue Cupid
E. lacturnus Godart, 1824 Indian Cupid
Cupido Schrank, 1801
C. alainus Staudinger, 1887 Staudinger's Cupid
C. buddhista Alpheraky, 1881 Shandur Cupid
Iolana Bethune-Baker, 1914
I. gigantea Grum-Grshimailo, 1885 Gilgit Mountain Blue
Bothrinia Chapman, 1909
B. chennelli de Niceville, 1884 Hedge Cupid
Tongeia Tutt, 1908
T. kala de Niceville, 1890 Black Cupid
T. pseudozuthus Huang, 2001 False Tibetan Cupid
Shijimia Matsumura, 1919
S. moorei Leech, 1889 Moore's Cupid
Talicada Moore, 1881
T. nyseus Guerin-Meneville, 1843
Pithecops Horsfield, 1828
P. fulgens Doherty, 1889 Blue Quaker
P. corvus Fruhstorfer, 1919 Forest Quaker
Azanus Moore, 1881
A. jesous Guerin-Meneville, 1849 African Babul Blue
A. ubaldus Stoll, 1782 Bright Babul Blue
A. uranus Butler, 1886 Dull Babul Blue
Neopithecops Distant, 1884
N. zalmora Butler, 1870 Quaker
Megisba Moore, 1881
M. malaya Horsfield, 1828 Malayan
Celastrina Tutt, 1906
C. argiolus Linnaeus, 1758 Hill Hedge Blue
C. gigas Hemming, 1928 Silvery Hedge Blue
C. hersilia Leech, 1893 Mishmi Hedge Blue
C. hugelii Moore, 1882 Large Hedge Blue
C. lavendularis Moore, 1877 Plain Hedge Blue
C. oreas Leech, 1893 Khasi Hedge Blue
Lestranicus Eliot & Kawazoe, 1983
L. transpectus Moore, 1879 White-banded Hedge Blue
Celatoxia Eliot & Kawazoe, 1983
C. albidisca Moore, 1884 Whitedisc Hedge Blue
C. marginata de Niceville, 1884 Margined Hedge Blue
Notarthrinus Chapman, 1908
N. binghami Chapman, 1908 Chapman's Hedge Blue
Acytolepis Toxopeus, 1927
A. lilacea Hampson, 1889 Hampson's Hedge Blue
A. puspa Horsfield, 1828 Common Hedge Blue
Oreolyce Toxopeus, 1927
O. dohertyi Tytler, 1915 Naga Hedge Blue
O. vardhana Moore, 1875 Dusky Hedge Blue
Callenya Eliot & Kawazoe, 1983
C. melaena Doherty, 1889 Metallic Hedge Blue
Monodontides Toxopeus, 1927
M. musina Snellen, 1892 Swinhoe's Hedge Blue
Udara Toxopeus, 1928
U. akasa Horsfield, 1828 White Hedge Blue
U. albocaerulea Moore, 1879 Albocerulean
U. dilecta Moore, 1879 Pale Hedge Blue
U. placidula Druce, 1895 Narrow-bordered Hedge Blue
U. selma Druce, 1895 Bicoloured Hedge Blue
U. singalensis R. Felder, 1868 Singhalese Hedge Blue
Euchrysops Butler, 1900
E. cnejus Fabricius, 1798 Gram Blue
Freyeria Courvoisier, 1920
F. putli Kollar, 1844 Small Grass Jewel
F. trochylus Freyer, 1845 Grass Jewel
Luthrodes Druce, 1895
L. pandava Horsfield, 1829 Plains Cupid
Chilades Moore, 1881
C. lajus Stoll, 1780 Lime Blue
C. parrhasius Fabricius, 1793 Small Cupid
Turanana Bethune-Baker, 1916
T. chitrali Charmeux & Pages, 2004 Chitral Argus Blue
Pseudophilotes Beuret, 1958
P. vicrama Moore, 1865 Eastern Baton Blue
Phengaris Doherty, 1891
P. atroguttata Oberthur, 1876 Great Spotted Blue
Plebejus Kluk, 1780
P. eversmanni Staudinger, 1886 Tibetan Jewel Blue
P. samudra Moore, 1875 Ladakh Jewel Blue
Aricia Reichenbach, 1817
A. agestis Denis & Schiffermueller, 1775 Orange-bordered Argus
A. artaxerxes Fabricius, 1793 Northern Brown Argus
A. astorica Evans, 1925 Astor Argus
Eumedonia Forster, 1938
E. eumedon Esper, 1780 Streaked Argus
Agriades Huebner, 1819
A. jaloka Moore, 1875 Greenish Mountain Blue
A. morsheadi Evans, 1923 Evans' Mountain Blue
A. pheretiades Eversmann, 1843 Tien Shan Blue
Albulina Tutt, 1909
A. asiatica Elwes, 1882 Azure Mountain Blue
A. chitralensis Tytler,1926 Chitral Green Underwing

A. chrysopis Grum-Grshimailo, 1888 Golden Green Underwing
A. galathea Blanchard, 1844 Large Green Underwing
A. lehanus Moore, 1878 Common Mountain Blue
A. metallica C. & R. Felder, 1865 Small Green Underwing
A. omphisa Moore, 1875 Dusky Green Underwing
A. pharis Fawcett, 1903 Fawcett's Mountain Blue
A. sikkima Moore, 1884 Sikkim Mountain Blue
Patricius Balint, 1992
P. younghusbandi Elwes, 1906 Chumbi Green Underwing
Plebejidea Koçak 1983
P. loewii Zeller, 1847 Large Jewel Blue
Kretania Beuret, 1959
Kretania beani Balint & Johnson, 1997 Bean's Jewel Blue
Farsia Zhdanko, 1992
F. ashretha Evans, 1925 Evans' Argus Blue
F. hanna Evans, 1932 Jewel Argus Blue
Alpherakya Zhdanko, 1994
A. devanica Moore, 1875 Dusky Meadow Blue
A. sarta Alpheraky, 1881 Brilliant Meadow Blue
Polyommatus Latreille, 1804
P. ariana Moore, 1865 Lahaul Meadow Blue
P. drasula Swinhoe, 1910 Ladakh Meadow Blue
Polyommatus dux Riley, 1926 Kumaon Meadow Blue
P. erigone Grum-Grshimailo, 1890 Grum-Grshimailo's Meadow Blue
P. hunza Grum-Grshimailo, 1890 Hunza Meadow Blue
P. icadius Grum-Grshimailo, 1890 Gilgit Meadow Blue
P. janetae Evans, 1927 Janet's Meadow Blue
P. pseuderos Moore, 1879 Kashmir Meadow Blue
P. pulchellus Bernardi, 1951 Bernardi's Meadow Blue
P. stoliczkanus C. & R. Felder, 1865 Stoliczka's Meadow Blue
Nymphalidae
Danainae
Danaini
Danaus Kluk, 1780
D. affinis Fabricius, 1775 Malay Tiger
D. chrysippus Linnaeus, 1758 Plain Tiger
D. genutia Cramer, 1779 Common Tiger
D. melanippus Cramer, 1777 White Tiger
Ideopsis Horsfield, 1829
I. juventa Cramer, 1777 Grey Glassy Tiger
I. similis Linnaeus, 1758 Blue Glassy Tiger
Parantica Moore, 1880
P. aglea Stoll, 1782 Glassy Tiger
P. agleoides C. & R. Felder, 1860 Dark Glassy Tiger
P. melaneus Cramer, 1775 Chocolate Tiger
P. swinhoei Moore, 1883 Swinhoe's Chocolate Tiger
P. nilgiriensis Moore, 1877 Nilgiri Tiger
P. pedonga Fujioka, 1970 Talbot's Chestnut Tiger
P. sita Kollar, 1844 Chestnut Tiger
Tirumala Moore, 1880
T. gautama Moore, 1877 Scarce Blue Tiger
T. limniace Cramer, 1775 Blue Tiger
T. septentrionis Butler, 1874 Dark Blue Tiger
Euploeini
Idea Fabricius, 1807
I. agamarschana C. & R. Felder, 1865 Burmese Tree Nymph
I. malabarica Moore, 1877 Malabar Tree Nymph
Euploea Fabricius, 1807
E. algea Godart, 1819 Long-branded Blue Crow
E. core Cramer, 1780 Common Crow
E. crameri Lucas, 1853 Spotted Black Crow
E. doubledayi C. & R. Felder, 1865
E. eunice Godart, 1819 Blue-branded King Crow
E. klugii Moore, 1858 King Crow
E. midamus Linnaeus, 1758 Spotted Blue Crow
E. mulciber Cramer, 1777 Striped Blue Crow
E. phaenareta Schaller, 1785 Great Crow
E. radamanthus Fabricius, 1793 Magpie Crow
E. sylvester Fabricius, 1793 Double-branded Crow

CALINAGINAE
Calinaga Moore, 1857
C. buddha Moore 1857 Freak

CHARAXINAE
Prothoini
Prothoe Huebner, 1824
P. franck Godart, 1824 Blue Begum
Charaxini
Polyura Billberg, 1820
P. agraria Swinhoe, 1887 Anomalous Nawab
P. arja C. & R. Felder, 1867 Pallid Nawab
P. athamas Drury, 1773 Common Nawab
P. bharata C. & R. Felder, 1867 Cryptic Nawab
P. delphis Doubleday, 1843 Jewelled Nawab
P. dolon Westwood, 1847 Stately Nawab
P. eudamippus Doubleday, 1843 Great Nawab
P. moori Distant, 1883 Malayan Nawab
P. narcaeus Hewitson, 1854 China Nawab
P. schreiber Godart, 1824 Blue Nawab
Charaxes Ochsenheimer, 1816
Ch. aristogiton C. & R. Felder, 1867 Scarce Tawny Rajah
Ch. bernardus Fabricius, 1793 Tawny Rajah
Ch. durnfordi Distant, 1884 Chestnut Rajah
Ch. kahruba Moore, 1895 Variegated Rajah
Ch. marmax Westwood, 1847 Yellow Rajah
Ch. psaphon Westwood, 1847 Plain Tawny Rajah
Ch. solon Fabricius, 1793 Black Rajah

AMATHUSINAE
Faunidini
Faunis Huebner, 1819
F. eumeus Drury, 1773 Large Faun
F. canens Huebner, 1826 Common Faun
Aemona Hewitson, 1868
A. amathusia Hewitson, 1867 Yellow Dryad
Stichophthalma C. & R. Felder, 1862
S. camadeva Westwood, 1848 Northern Junglequeen
S. nourmahal Westwood, 1851 Chocolate Junglequeen
S. sparta de Niceville, 1894 Manipur Junglequeen
Amathusia Fabricius, 1807
A. andamanensis Fruhstorfer, 1899 Andaman Palmking
A. phidippus Linnaeus, 1763 Common Palmking
Amathuxidia Staudinger, 1887
A. amythaon Doubleday, 1847 Koh-i-Noor
Thaumantis Huebner, 1826
T. diores Doubleday, 1845 Jungleglory
Thauria Moore, 1894
T. lathyi Fruhstorfer, 1902 Jungleking
Discophora Boisduval, 1836
D. deo de Niceville, 1898 Banded Duffer
D. lepida Moore, 1857 Southern Duffer
D. sondaica Boisduval, 1836 Common Duffer
D. timora Westwood, 1850 Great Duffer
Enispe Doubleday, 1848
E. cycnus Westwood, 1851 Blue Caliph
E. euthymius Doubleday, 1845 Red Caliph
E. intermedia Rothschild, 1916 Malayan Red Caliph

SATYRINAE
Elymniini
Elymnias Huebner, 1818

E. cottonis Hewitson, 1874 Andaman Palmfly
E. hypermnestra Linnaeus, 1763 Common Palmfly
E. malelas Hewitson, 1863 Spotted Palmfly
E. nesaea Linnaeus, 1764 Tiger Palmfly
E. panthera Fabricius, 1787 Nicobar Palmfly
E. patna Westwood, 1851 Blue-striped Palmfly
E. peali Wood-Mason, 1883 Peal's Palmfly
E. penaga Westwood, 1851 Pointed Palmfly
E. vasudeva Moore, 1857 Jezabel Palmfly
Zetherini
Neorina Westwood, 1850
N. hilda Westwood, 1850 Yellow Owl
N. patria Leech, 1891 White Owl
Penthema Doubleday, 1848
P. lisarda Doubleday, 1845 Yellow Kaiser
Ethope Moore, 1866
E. himachala Moore, 1857 Dusky Diadem
Melanitini
Melanitis Fabricius, 1807
M. leda Linnaeus, 1758 Common Evening Brown
M. phedima Cramer, 1780 Dark Evening Brown
M. zitenius Herbst, 1796 Great Evening Brown
Cyllogenes Butler, 1868
C. janetae de Niceville, 1887 Scarce Evening Brown
C. suradeva Moore, 1857 Branded Evening Brown
Parantirrhoea Wood-Mason, 1881
P. marshalli Wood-Mason, 1881 Travancore Evening Brown
Satyrini
Lethe Huebner, 1819
L. andersoni Atkinson, 1871 Anderson's Silverstripe
L. atkinsonia Hewitson, 1876 Small Goldenfork
L. baladeva Moore, 1866 Treble Silverstripe
L. bhairava Moore, 1857 Rusty Forester
L. brisanda de Niceville, 1886 Dark Forester
L. chandica Moore, 1858 Angled Red Forester
L. confusa Aurivillius, 1898 Banded Treebrown
L. dakwania Tytler, 1939 Garhwal Woodbrown
L. dura Marshall, 1882 Scarce Lilacfork
L. distans Butler, 1870 Scarce Red Forester
L. drypetis Hewitson, 1863 Tamil Treebrown
L. europa Fabricius, 1775 Bamboo Treebrown
L. goalpara Moore, 1866 Large Goldenfork
L. gemina Leech, 1891Tytler's Treebrown
L. gulnihal de Niceville, 1887 Dull Forester
L. isana Kollar, 1844 Common Forester
L. jalaurida de Niceville, 1880 Small Silverfork
L. kabrua Tytler, 1914 Manipur Goldenfork
L. kanjupkula Tytler, 1914 Manipur Woodbrown
L. kansa Moore, 1857 Bamboo Forester
L. latiaris Hewitson, 1862 Pale Forester
L. maitrya de Niceville, 1881 Barred Woodbrown
L. margaritae Elwes, 1882 Bhutan Treebrown
L. mekara Moore, 1858 Common Red Forester
L. moelleri Elwes, 1887 Moeller's Silverfork
L. naga Doherty, 1889 Naga Treebrown
L. nicetas Hewitson, 1863 Yellow Woodbrown
L. nicetella de Niceville, 1887 Small Woodbrown
L. ramadeva de Niceville, 1887 Single Silverstripe
L. rohria Fabricius, 1787 Common Treebrown
L. satyavati de Niceville, 1880 Pallid Forester
L. scanda Moore, 1857 Blue Forester
L. serbonis Hewitson, 1876 Brown Forester
L. siderea Marshall, 1881 Scarce Woodbrown
L. sidonis Hewitson, 1863 Common Woodbrown
L. sinorix Hewitson, 1863 Tailed Red Forester
L. sura Doubleday, 1849 Scarce Lilacfork
L. tristigmata Elwes, 1887 Spotted Mystic
L. verma Kollar, 1844 Straight-banded Treebrown
L. vindhya C. & R. Felder, 1859 Black Forester
L. visrava Moore, 1866 White-edged Woodbrown
Neope Moore, 1866
N. armandii Oberthur, 1876 Chinese Labyrinth
N. bhadra Moore, 1857 Tailed Labyrinth
N. pulaha Moore, 1858 Veined Labyrinth
N. pulahina Evans, 1923 Scarce Labyrinth
N. pulahoides Moore, 1892 Assam Veined Labyrinth
N. yama Moore, 1858 Dusky Labyrinth
Lasiommata Westwood, 1841
L. maerula C. & R. Felder, 1867 Scarce Wall
L. menava Moore, 1865 Dark Wall
L. schakra Kollar, 1844 Common Wall
Kirinia Moore, 1893
K. eversmanni Eversmann, 1847 Yellow Wall
Chonala Moore, 1893
Ch. masoni Elwes, 1882 Chumbi Wall
Rhaphicera Butler, 1867
R. moorei Butler, 1867 Small Tawny Wall
R. satricus Doubleday, 1849 Large Tawny Wall
Orinoma Gray, 1846
O. damaris Gray, 1846 Tigerbrown
Heteropsis Westwood, 1850
H. adolphei Guerin-Meneville, 1843 Redeye Bushbrown
H. malsara Moore, 1857 White-line Bushbrown
H. mamerta Stoll, 1780 Blind-eye Bushbrown
Mycalesis Huebner, 1818
M. adamsoni Watson, 1897 Watson's Bushbrown
M. anapita Moore, 1858 Tawny Bushbrown
M. anaxias Hewitson, 1862 Whitebar Bushbrown
M. annamitica Fruhstorfer, 1906 Blindeyed Bushbrown
M. evansii Tytler, 1914 Tytler's Bushbrown
M. francisca Stoll, 1780 Lilacine Bushbrown
M. gotama Moore, 1857 Chinese Bushbrown
M. heri Moore, 1857 Moore's Bushbrown
M. igilia Fruhstorfer, 1911 Small Long-brand Bushbrown
M. intermedia Moore, 1892 Intermediate Bushbrown
M. lepcha Moore, 1880 Lepcha Bushbrown
M. malsarida Butler, 1868 Plain Bushbrown
M. manii Doherty, 1886 Nicobar Bushbrown
M. mestra Hewitson, 1862 White-edged Bushbrown
M. mineus Linnaeus, 1758 Dark-brand Bushbrown
M. misenus de Niceville, 1889 De Niceville's Bushbrown
M. mnasicles Hewitson, 1864 Cyclops Bushbrown
M. mystes de Niceville, 1891 Many-tufted Bushbrown
M. nicotia Westwood, 1850 Brighteye Bushbrown
M. oculus Marshall, 1881 Red-disc Bushbrown
M. orcha Evans, 1920 Pale-brand Bushbrown
M. orseis Hewitson, 1864 Purple Bushbrown
M. patiana Eliot, 1969 Eliot's Bushbrown
M. patnia Moore, 1857 Glad-eye Bushbrown
M. perseus Fabricius, 1775 Common Bushbrown
M. radza Moore, 1877 Andaman Bushbrown
M. suaveolens Wood-Mason & de Niceville, 1883 Wood-Mason's Bushbrown
M. visala Moore, 1858 Long-brand Bushbrown
OrsotriaenaWallengren, 1858
O. medus Fabricius, 1775 Nigger
Zipaetis Hewitson, 1863
Z. saitis Hewitson, 1863 Tamil Catseye
Z. scylax Hewitson, 1863 Dark Catseye
Erites Westwood, 1851
E. falcipennis Wood-Mason & de Niceville, 1883 Common Cyclops
Coelites Westwood, 1850
C. nothis Westwood, 1850 Scarce Catseye

Ragadia Westwood, 1851
R. crisilda Hewitson, 1862 Striped Ringlet
Hyponephele Muschamp, 1915
H. carbonelli Lukhtanov, 1995 Baltistan Meadowbrown
H. cheena Moore, 1865 Branded Meadowbrown
H. coenonympha C. & R. Felder, 1867 Spotted Meadowbrown
H. davendra Moore, 1865 White-ringed Meadowbrown
H. brevistigma Moore, 1893 Short-branded Meadowbrown
H. tenuistigma Moore, 1893 Lesser White-ringed Meadowbrown
H. pulchella C. & R. Felder, 1867 Tawny Meadowbrown
H. pulchra C. & R. Felder, 1867 Dusky Meadowbrown
H. hilaris Staudinger, 1886 Pamir Meadowbrown
Callerebia Butler, 1867
C. annada Moore, 1858 Ringed Argus
C. baileyi South, 1913 White-bordered Argus
C. dibangensis Roy, 2013 Roy's Argus
C. hybrida Butler, 1880 Hybrid Argus
C. nirmala Moore, 1865 Common Argus
C. orixa Moore, 1872 Moore's Argus
C. scanda Kollar, 1844 Pallid Argus
C. suroia Tytler, 1914 Manipur Argus
Paralasa Moore, 1893
P. chitralica Evans, 1923 Chitral Argus
P. kalinda Moore, 1865 Scarce Mountain Argus
P. mani de Niceville, 1881 Yellow Argus
P. shallada Lang, 1881 Mountain Argus
Loxerebia Watkins, 1925
L. narasingha Moore, 1857 Mottled Argus
Ypthima Huebner, 1818
Y. affectata Elwes & Edwards, 1893 Khasi Fivering
Y. asterope Klug, 1832 Common Threering
Y. atra Cantlie & Norman, 1959 Black Fivering
Y. avanta Moore, 1875 Jewel Fourring
Y. baldus Fabricius, 1775 Common Fivering
Y. bolanica Marshall, 1882 Desert Fourring
Y. cantliei Norman, 1958 Cantlie's Fourring
Y. ceylonica Hewitson, 1865 White Fourring
Y. chenu Guerin-Meneville, 1843 Nilgiri Fourring
Y. davidsoni Eliot, 1967 Davidson's Fivering
Y. dohertyi Moore, 1893 Great Fivering
Y. fusca Elwes & Edwards, 1893 Assam Threering
Y. hannyngtoni Eliot, 1967 Hannyngton's Fivering
Y. huebneri Kirby, 1871 Common Fourring
Y. hyagriva Moore, 1857 Brown Argus
Y. indecora Moore, 1882 Western Fivering
Y. inica Hewitson, 1865 Lesser Threering
Y. kasmira Moore, 1884 Kashmir Fourring
Y. lisandra Cramer, 1780 Jewel Fourring
Y. lycus de Niceville, 1889 Plain Threering
Y. methora Hewitson, 1865 Variegated Fivering
Y. nareda Kollar, 1844 Large Threering
Y. newara Moore, 1875 Newar Threering
Y. nikaea Moore, 1875 Moore's Fivering
Y. norma Westwood, 1851 Burmese Threering
Y. parasakra Eliot, 1987 Himalayan Fourring
Y. persimilis Elwes & Edwards, 1893 Manipur Fivering
Y. philomela Linnaeus, 1763 Baby Fivering
Y. sakra Moore, 1857 Himalayan Fivering
Y. savara Grose-Smith, 1887 Pallid Fivering
Y. similis Elwes & Edwards, 1893 Eastern Fivering
Y. singala R. Felder, 1868 Small Jewel Fourring
Y. sobrina Elwes & Edwards, 1893 Karen Fivering
Y. striata Hampson, 1888 Nilgiri Jewel Fourring
Y. watsoni Moore, 1893 Looped Threering
Y. yphthimoides Moore, 1881 Palni Fourring
Oeneis Huebner, 1819
O. buddha Grum-Grshimailo, 1891 Tibetan Satyr
Paroeneis Moore, 1893
P. pumilus C. & R. Felder, 1867 Mountain Satyr
P. sikkimensis Staudinger, 1889 Sikkim Satyr
Karanasa Moore, 1893
K. alpherakyi Avinoff, 1910 Avinoff's Satyr
K. bolorica Grum-Grshimailo, 1888 Turkestan Satyr
K. cadesia Moore, 1875 Moore's Satyr
K. huebneri C. & R. Felder, 1867 Tawny Satyr
K. modesta Moore, 1893 Small Satyr
K. moorei Evans, 1912 Shandur Satyr
K. leechi Grum-Grshimailo, 1890 Leech's Satyr
K. rohtanga Avinoff & Sweadner, 1951 Rohtang Satyr
Satyrus Latreille, 1810
S. pimpla C. & R. Felder, 1867 Black Satyr
Aulocera Butler, 1867
A. brahminus Blanchard, 1853 Narrow-banded Satyr
A. loha Doherty, 1886 Doherty's Satyr
A. padma Kollar, 1844 Great Satyr
A. saraswati Kollar, 1844 Striated Satyr
A. swaha Kollar, 1844 Common Satyr
Hipparchia Fabricius, 1807
H. parisatis Kollar, 1849 White-edged Rockbrown
Chazara Moore, 1893
C. heydenreichi Lederer, 1853 Shandur Rockbrown
Pseudochazara de Lesse, 1951
P. baldiva Moore, 1865 Kashmir Rockbrown
P. droshica Tytler, 1926 Tytler's Rockbrown
Kanetisa Moore, 1893
K. digna Marshall, 1883 Chitrali Satyr

LIMENITIDINAE
Tribe **Limenitidini**
Neptis Fabricius, 1807
N. ananta Moore, 1858 Yellow Sailer
N. armandia Oberthur, 1876 Variegated Sailer
N. capnodes Fruhstorfer, 1908 Eliot's Sailer
N. cartica Moore, 1872 Plain Sailer
N. clinia Moore, 1872 Clear Sailer
N. cydippe Leech, 1890 Chinese Yellow Sailer
N. harita Moore, 1875 Dingiest Sailer
N. hylas Linnaeus, 1758 Common Sailer
N. ilira Kheil, 1884 Dark Dingy Sailer
N. jumbah Moore, 1858 Chestnut-streaked Sailer
N. magadha C. & R. Felder, 1867 Spotted Sailer
N. mahendra Moore, 1872 Himalayan Sailer
N. manasa Moore, 1858 Pale Hockeystick Sailer
N. miah Moore, 1857 Small Yellow Sailer
N. namba Tytler, 1915 Manipur Yellow Sailer
N. narayana Moore, 1858 Broadstick Sailer
N. nashona Swinhoe, 1896 Less Rich Sailer
N. nata Moore, 1858 Sullied Sailer
N. nemorum Oberthur, 1906 Naga Hockeystick Sailer
N. nycteus de Niceville, 1890 Hockeystick Sailer
N. pseudovikasi Moore, 1899 Dingy Sailer
N. radha Moore, 1857 Great Yellow Sailer
N. sankara Kollar, 1844 Broad-banded Sailer
N. sappho Pallas, 1771 Pallas' Sailer
N. soma Moore, 1858 Creamy Sailer
N. zaida Doubleday, 1848 Pale Green Sailer
Phaedyma C. Felder, 1861
Ph. aspasia Leech, 1890 Great Hockeystick Sailer
Ph. columella Cramer, 1780 Short-banded Sailer
Lasippa Moore, 1898
L. tiga Moore, 1858 Burmese Lascar
L. viraja Moore, 1872 Yellowjack Sailer

Pantoporia Huebner, 1819
P. assamica Moore, 1881 Assamese Lascar
P. aurelia Staudinger, 1886 Baby Lascar
P. bieti Oberthur, 1894 Tytler's Lascar
P. hordonia Stoll, 1790 Common Lascar
P. paraka Butler, 1879 Perak Lascar
P. sandaka Butler, 1892 Extra Lascar
Athyma Westwood, 1850
A. asura Moore, 1858 Studded Sergeant
A. cama Moore, 1858 Orange Staff Sergeant
A. nefte Cramer, 1780 Colour Sergeant
A. jina Moore, 1858 Bhutan Sergeant
A. kanwa Moore, 1858 Dot-dash Sergeant
A. larymna Doubleday, 1848 Great Sergeant
A. opalina Kollar, 1844 Himalayan Sergeant
A. orientalis Elwes, 1888 Oriental Sergeant
A. perius Linnaeus, 1758 Common Sergeant
A. pravara Moore, 1858 Unbroken Sergeant
A. ranga Moore, 1858 Blackvein Sergeant
A. reta Moore, 1858 Malay Staff Sergeant
A. rufula de Niceville, 1889 Andaman Sergeant
A. selenophora Kollar, 1844 Staff Sergeant
A. whitei Tytler, 1940 Cryptic Sergeant
A. zeroca Moore, 1872 Small Staff Sergeant
Limenitis Fabricius, 1807
L lepechini Erschoff, 1874 Chitral White Admiral
L. trivena Moore, 1864 Indian White Admiral
Moduza Moore, 1881
M. procris Cramer, 1777 Commander
Parasarpa Moore, 1898
P. dudu Doubleday, 1848 White Commodore
P. zayla Doubleday, 1848 Bicolour Commodore
Sumalia Moore, 1898
S. daraxa Doubleday, 1848 Green Commodore
S. zulema Doubleday, 1848 Scarce White Commodore
Auzakia Moore, 1898
A. danava Moore, 1858 Commodore
Bhagadatta Moore, 1898
B. austenia Moore, 1872 Grey Commodore
Lebadea C. Felder, 1861
L. martha Fabricius, 1787 Knight
Parthenos Huebner, 1819
P. sylla Donovan, 1842 Clipper
Neurosigma Butler, 1869
N. siva Westwood, 1850 Panther
Adoliadini
Abrota Moore, 1857
A. ganga Moore, 1857 Sergeant-major
Cynitia Snellen, 1895
C. cocytus Fabricius, 1787 Lavender Count
C. lepidea Butler, 1868 Grey Count
C. telchinia Menetries, 1857 Blue Baron
Tanaecia Butler, 1869
T. cibaritis Hewitson, 1874 Andaman Viscount
T. jahnu Moore, 1858 Plain Earl
T. julii Lesson, 1837 Common Earl
Euthalia Huebner, 1819
E. aconthea Cramer, 1777 Common Baron
E. alpheda Godart, 1824 Streaked Baron
E. anosia Moore, 1858 Grey Baron
E. confucius Westwood, 1850 Chinese Duke
E. curvifascia Tytler, 1915 Naga Duke
E. durga Moore, 1858 Blue Duke
E. duda Staudinger, 1886 Blue Duchess
E. evelina Stoll, 1790 Red-spot Duke
E. eriphylae de Niceville, 1891 White-tipped Baron
E. franciae Gray, 1846 French Duke
E. iva Moore, 1858 Grand Duke
E. lengba Tytler, 1940 Tytler's Duchess
E. lubentina Cramer, 1777 Gaudy Baron
E. malaccana Fruhstorfer, 1899 Fruhstorfer's Baron
E. merta Moore, 1859 Dark Baron
E. monina Fabricius, 1787 Powdered Baron
E. nara Moore, 1859 Bronze Duke
E. narayana Grose-Smith & Kirby, 1891 Burmese Baron
E. patala Kollar, 1844 Grand Duchess
E. phemius Doubleday, 1848 White-edged Blue Baron
E. recta de Niceville, 1886 Redtail Maquis
E. sahadeva Moore, 1859 Green Duke
E. saitaphernes Fruhstorfer, 1913 Mottled Baron
E. teuta Doubleday, 1848 Banded Marquis
E. thawgawa Tytler, 1940 Tytler's Duke
Symphaedra Huebner, 1818
S. nais Forster, 1771 Baronet
Lexias Boisduval, 1832
L. cyanipardus Butler, 1869 Great Archduke
L. dirtea Fabricius, 1793 Dark Archduke
L. pardalis Moore, 1878 Archduke

Subfamily **HELICONIINAE**
Tribe **Argynnini**
Argynnis Fabricius, 1807
A. aglaja Linnaeus, 1758 Dark Green Silverspot
A. childreni Gray, 1831 Great Silverstripe
A. clara Blanchard, 1844 Silverstreak
A. hyperbius Linnaeus, 1763 Indian Fritillary
A. jainadeva Moore, 1864 Highbrown Silverspot
A. kamala Moore, 1857 Common Silverstripe
A. laodice Pallas, 1771 Eastern Silverstripe
A. pandora Schiffermuller 1775 Cardinal
Issoria Huebner, 1819
I. altissima Elwes, 1882 Mountain Silverspot
I. gemmata Butler, 1881 Gem Silverspot
I. lathonia Linnaeus, 1758 Queen of Spain Fritillary
I. mackinnoni de Niceville, 1891 Brilliant Silverspot
Boloria Moore, 1900
B. erubescens Staudinger, 1901 Whitespot Fritillary
B. generator Staudinger, 1886 Hunza Silverspot
B. jerdoni Lang, 1868 Jerdon's Silverspot
B. sipora Moore, 1875 Straightwing Silverspot
Heliconiini
Phalanta Horsfield, 1829
P. alcippe Stoll, 1782 Small Leopard
P. phalantha Drury, 1773 Common Leopard
Cupha Billberg, 1820
C. erymanthis Drury, 1773 Rustic
Vagrans Hemming, 1934
V. egista Cramer, 1780 Vagrant
Vindula Hemming, 1934
V. erota Fabricius, 1793 Cruiser
Algia Herrich-Schaeffer, 1864
A. fasciata C. & R. Felder, 1860 Branded Yeoman
Cirrochroa Doubleday, 1847
C. aoris Doubleday, 1847 Large Yeoman
C. nicobarica Wood-Mason & de Nicceville, 1881 Nicobar Yeoman
C. thais Fabricius, 1787 Tamil Yeoman
C. tyche C. & R. Felder, 1861 Common Yeoman

BIBLIDINAE
Biblidini
Ariadne Horsfield, 1829
A. ariadne Linnaeus, 1763 Angled Castor
A. merione Cramer, 1777 Common Castor

Laringa Moore, 1901
L. horsfieldi Boisduval, 1833 Banded Dandy
Byblia Huebner, 1819
B. ilithyia Drury, 1773 Joker

APATURINAE
Apaturini
Rohana Moore, 1880
R. parisatis Westwood, 1850 Black Prince
R. parvata Moore, 1857 Brown Prince
Eulaceura Butler, 1872
E. manipurensis Tytler, 1915 Tytler's Emperor
Chitoria Moore, 1896
C. naga Tytler, 1915 Naga Emperor
C. sordida Moore, 1866 Sordid Emperor
C. ulupi Doherty, 1889 Tawny Emperor
Mimathyma Moore, 1896
M. ambica Kollar, 1844 Indian Purple Emperor
M. bhavana Moore, 1881 Bhutan Emperor
M. chevana Moore, 1866 Sergeant Emperor
M. chitralensis Evans, 1912 Chitral Emperor
Dilipa Moore, 1857
D. morgiana Westwood, 1850 Golden Emperor
Sephisa Moore, 1882
S. chandra Moore, 1858 Eastern Courtier
S. dichroa Kollar, 1844 Western Courtier
Helcyra C. Felder, 1860
H. hemina Hewitson, 1864 White Emperor
Herona Doubleday, 1848
H. marathus Doubleday, 1848 Pasha
Euripus Doubleday, 1848
E. consimilis Westwood, 1850 Painted Courtesan
E. nyctelius Doubleday, 1845 Courtesan
Hestina Westwood, 1850
H. nicevillei Moore, 1895 Scarce Siren
H. persimilis Westwood, 1850 Siren
Hestinalis Bryk, 1938
H. nama Doubleday, 1844 Circe
Sasakia Moore, 1896
S. funebris Leech, 1891 Empress

CYRESTINAE
Cyrestini
Cyrestis Boisduval, 1832
C. cocles Fabricius, 1787 Marbled Map
C. tabula de Niceville, 1883 Nicobar Map
C. thyodamas Boisduval, 1846 Common Map
Chersonesia Distant, 1883
C. intermedia Martin, 1895 Wavy Maplet
C. risa Doubleday, 1848 Common Maplet
Pseudergolini
Pseudergolis C. & R. Felder, 1867
P. wedah Kollar, 1848 Tabby
Stibochiona Butler, 1869
S. nicea Gray, 1846 Popinjay
Dichorragia Butler, 1869
D. nesimachus Doyere, 1840 Constable

NYMPHALINAE
Melitaeini
Melitaea Fabricius, 1807
M. arcesia Bremer, 1861 Blackvein Fritillary
M. shandura Evans, 1924 Shandur Fritillary
M. fergana Staudinger, 1882 Uzbek Fritillary
M. nadezhdae Sheljuzhko 1912 Sheljuzhko's Fritillary
M. balbina Tytler, 1926 Pamir Fritillary
Nymphalini
Symbrenthia Huebner, 1819
S. brabira Moore, 1872 Himalayan Jester
S. doni Tytler, 1940 Naga Jester
S. hypselis Godart, 1824 Spotted Jester
S. lilaea Hewitson, 1864 Common Jester
S. niphanda Moore, 1872 Blue-tailed Jester
S. silana de Niceville, 1885 Scarce Jester
Araschnia Huebner, 1819
A. prorsoides Blanchard, 1871 Mongol
Nymphalis Kluk, 1780
N. xanthomelas Esper, 1781 Large Tortoiseshell
Aglais Dalman, 1816
A. caschmirensis Kollar, 1844 Indian Tortoiseshell
A. ladakensis Moore, 1878 Ladakh Tortoiseshell
A. rizana Moore, 1872 Mountain Tortoiseshell
Kaniska Moore, 1899
K. canace Linnaeus, 1763 Blue Admiral
Polygonia Huebner, 1819
P. c-album Linnaeus, 1758 Comma
P. l-album Esper, 1781 False Comma
P. undina Grum-Grshimailo, 1890 Pamir Comma
Vanessa Fabricius, 1807
V. cardui Linnaeus, 1758 Painted Lady
V. indica Herbst, 1794 Indian Red Admiral
Junoniini
Junonia Huebner, 1819
J. almana Linnaeus, 1758 Peacock Pansy
J. atlites Linnaeus, 1763 Grey Pansy
J. hierta Fabricius, 1798 Yellow Pansy
J. iphita Cramer, 1779 Chocolate Pansy
J. lemonias Linnaeus, 1758 Lemon Pansy
J. orithya Linnaeus, 1758 Blue Pansy
Kallimini
Hypolimnas Huebner, 1819
H. anomala Wallace, 1869 Malayan Eggfly
H. bolina Linnaeus, 1758 Great Eggfly
H. misippus Linnaeus, 1764 Danaid Eggfly
Kallima Doubleday, 1849
K. albofasciata Moore, 1877 White Oakleaf
K. horsfieldi Kollar, 1844 Southern Blue Oakleaf
K. inachus Boisduval, 1846 Orange Oakleaf
K. knyvetti de Niceville, 1886 Scarce Blue Oakleaf
Doleschallia C. & R. Felder, 1860
D. bisaltide Cramer, 1777 Autumn Leaf
Rhinopalpa C. & R. Felder, 1860
R. polynice Cramer, 1799 Wizard
Yoma Doherty, 1886
Y. sabina Cramer, 1780 Lurcher

ACRAEINAE
Acraeini
Acraea Fabricius, 1807
A. issoria Huebner, 1819 Yellow Coster
A. violae Fabricius, 1793 Tawny Coster
Cethosiini
Cethosia Fabriciuis, 1807
C. biblis Drury, 1773 Red Lacewing
C. cyane Drury, 1773 Leopard Lacewing
C. nietneri C. & R. Felder, 1867 Tamil Lacewing

LIBYTHEINAE
Libythea Fabricius, 1807
L. lepita Moore, 1858 Common Beak
L. myrrha Godart, 1819 Club Beak
L. narina Godart, 1819 White-spotted Beak

RESOURCES

Knowledgeable people in the regions covered by this book who can help with advice on where to see butterflies include:

Pakistan
M. Akram Awam
(pakistanbutterflies@gmail.com)

Nepal
Tashi Ghale (tashirghale@gmail.com)
Bhaiya Khanal (baya2000@live.com)

Bhutan
Sonam Dorji (bhutanreality@yahoo.com)

Bangladesh
Amit Kumar Neogi
(amit_jnu52@yahoo.com)
Tanvir Ahmed Shaikot (Tanvir Ahmed Shaikot2023jnu@gmail.com)

Sri Lanka
Tharaka Priyadarshana (tharakas.priyadarshana@gmail.com)
Ishara Harshajit
(ishararulzz777@gmail.com)

Buxa Tiger Reserve, West Bengal
Subhajit Mazumder
(mazumder.subhajit@gmail.com)

Sikkim
Matrika Sharma
(sharmamatrika1978@gmail.com)
Kusal Gurung (korongee@gmail.com)

Assam
Parixit Kafley (parixitk1@gmail.com),
Atanu Bora (atanubora47@yahoo.com)
Atanu Bose (atanusays@gmail.com)

Gujarat
Pratiksha Patel
(butterfliesofgujarat@gmail.com)

Western Himalaya
Peter Smetacek
(butterflyresearchcentre@gmail.com)

FURTHER READING

Bozano, G. C. (ed.) 1999. *Guide to the Butterflies of the Palaearctic Region*. More than 30 volumes. Omnes Artes, Milan.

Corbet, A. S. & H. M. Pendlebury. 1978. *The Butterflies of the Malay Peninsula*. Third edition revised by J. N. Eliot. 478 pp., 35 pl. 438 figs. Malayan Nature Society, Kuala Lumpur.

Scoble, M. J. 1992. *The Lepidoptera. Form, Function and Diversity*. xi+404 pp. British Museum (Natural History), London.

Smetacek, P. 2015. *Papilionid Butterflies of the Indian Subcontinent*. 121 pp. Butterfly Research Centre, Bhimtal and Indinov Publishing, New Delhi. (e-book, free download at https://www.researchgate.net/publication/272383021_Papilionid_Butterflies_of_the_Indian_Subcontinent).

Varshney, R. K. & Smetacek, P. (eds.) 2015. *A Synoptic Catalogue of the Butterflies of India*. Butterfly Research Centre, Bhimtal and Indinov Publishing, New Delhi. ii + 261 pp., 8 pl. (e-book, free download at https://www.researchgate.net/publication/287980260_A_Synoptic_Catalogue_of_the_Butterflies_of_India).